AF568142

TEXT BOOK
OF
ELEMENTARY CO-ORDINATE

DPH MATHEMATICS SERIES

TEXT BOOK OF ELEMENTARY CO-ORDINATE

By

A.K. Sharma

DISCOVERY PUBLISHING HOUSE
NEW DELHI-110002

First Published-2005

ISBN 81-8356-029-6

Published by

DISCOVERY PUBLISHING HOUSE

4831/24, Ansari Road, Prahlad Street,
Darya Ganj, New Delhi-110002 (India)
Phone: 23279245 • Fax: 91-11-23253475
E-mail:dphtemp@indiatimes.com

Printed at:

Amit Enterprises, Delhi

Preface

This book "Text Book of Elementary Co-ordinate" has been specially written to meet the requirements of B.A./B.Sc. students of all Indian Universities.

The subject matter has been discussed in such a simple way that the student will find no difficulty to understand it. The proof of various theorems and examples have been given with minute details. Each chapter of this book contains complete theory and large number of solved examples. Sufficient problems have also been selected form various Indian Universities and competitive examination.

I hope that this book warmly received by the student and teachers.

I wish to thank my colleague and friends for countless helpful remark and suggestions. The author wish to express his thanks to the publisher M/s Discovery Publishing House, New Delhi for bringing out this book in present nice form.

A.K. Sharma

Contents

1

ELLIPSE

Definition: The ellipse is a coni section in which the eccentricity e is less than unity. An ellipse is the locus of the point which moves in a plane so that its distance from a fixed point bears a constant ratio, less than unity, to its distance from a fixed straight line.

The fixed point is known as the focus. The fixed straight line is known as directrix of the ellipse and the constant ratio is called the eccentricity of the ellipse and is denoted by e.

EQUATION OF AN ELLIPSE

To find the equation of the ellipse in the form $\frac{x^2}{a^2}+\frac{y^2}{b^2}=1$

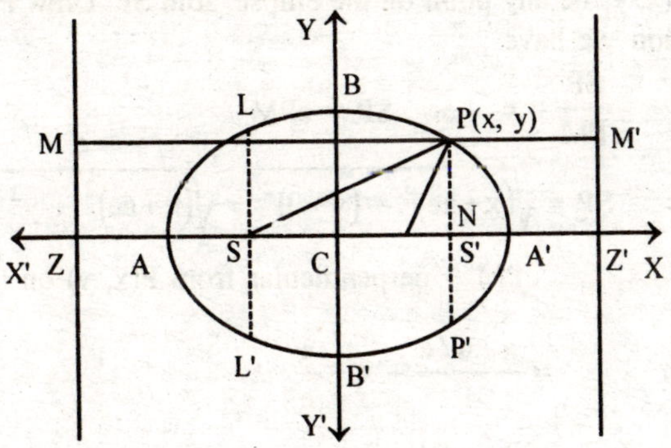

Proof:

There will be a point A on SZ, such that

$$SA = eAZ \qquad ...(i)$$

sin q e < 1, there will be another point A' on ZS produced such that

$$SA' = e\,A'Z \qquad ...(ii)$$

Thus, by definition, A and A' lie on the ellipse. Let C be the middle point of AA' and let AA' = 2a, then we have

$$CA = CA' = a$$

Adding (i) and (ii), we have

$$AS + A'S = e[ZA + ZA']$$

$$\Rightarrow \quad AA' = e(CZ - CA + CA' + CZ)$$

$$\Rightarrow \quad 2a = e(2CZ)$$

$$\Rightarrow \quad CZ = a/e. \qquad \text{...(iii)}$$

Again substracting (i) from (ii), we obtain

$$A'S - AS = e(ZA' - ZA)$$

$$\Rightarrow \quad (A'C + CS) - (AC - CS) = eAA'$$

$$2CS = e.\ 2a$$

$$\therefore \quad CS = ae$$

Take C at origin, AA' as a-axis and a line through C and perpendicular to AA' as y-axis.

$\therefore$ From (iii) and (iv), the co-ordinates of S are (–ae, 0) and the equation of directrix ZM is

$$x = -a/e \quad \Rightarrow \quad x + a/e = 0. \qquad \text{...(v)}$$

Let P(x, y) be any point on the ellipse. Join SP. Draw PM ⊥ ZM. Then by definition we have

$$\frac{SP}{PM} = e \quad \text{or} \quad SP = ePM \qquad \text{...(vi)}$$

where $SP = \sqrt{(x+ae)^3 + (y-0)^2} = \sqrt{(x+ae)^2 + y^2}$ ...(vii)

and PM = perpendicular from P(x, y) on (v)

$$= \frac{x + a/e}{1} + x + \frac{a}{e} \qquad \text{...(viii)}$$

$\therefore$ from (vi), (vii) and (viii), we get

$$\sqrt{(x+ae)^2 + y^2} = e\left(x + \frac{a}{e}\right) = ex + a$$

Squaring both sides, we obtain

$$x^2 + 2aex + a^2e^2 + y^2 = e^2x^2 + 2exa + a^2$$

$$\Rightarrow \quad x^2(1 - e^2) + y^2 = a^2(1 - e^2)$$

dividing both sides by $a^2(1 - e^2)$, we have

$$\frac{x^2}{a^2} + \frac{y^2}{a^2\left(1-e^2\right)} = 1.$$

Putting $a^2(1 - e^2) = b^2$, the required equation is

$$\frac{x^2}{a^2} + \frac{y^2}{b^2} = 1.$$

Note. The y-axis meets the ellipse $\frac{x^2}{a^2} + \frac{y^2}{b^2} = 1$, at the points

$$\frac{y^2}{b^2} = 1 \quad \text{or} \quad y = \pm b$$

i e. the y-axis meets the ellipse in two points B and B', such that

$$CB = CB' = b$$

Definition:

(i) The lines AA' and BB' are called *major* and *minor axis* respectively. Both together are called principal axes.

(ii) The point A and A', the extremities of the major axis are called *vertices* of the ellipse.

(iii) C is called the **centre** of the ellipse. The centre C bisects every chord which passes through it.

SECOND FOCUS AND SECOND DIRECTRIX OF AN ELLIPSE

Prove that the ellipse has a second focus and a second directrix.

Proof:

On the positive side of the origin take a point S' which is such that SC = CS' = ae and another point Z' such that

$$ZC = CZ' = a/e$$

Draw Z/K/ perpendicular to ZZ' and PM' perpendicular to Z'K'. The equation $\frac{x^2}{a^2} + \frac{y^2}{b^2} = 1$ in the form

$$x^2 - 2ax + a^2e^2 + y^2 = e^2x^2 - 2aex + a^2$$

$$\Rightarrow \quad (x - ae)^2 + y^2 = e^2\,(a/e - x)^2$$

$$\Rightarrow \quad S'P^2 + e^2 = PM'^2.$$

Hence any point P of the curve is such that its distance from S' is e times its distance from Z'K' so that we should have obtained the same curve. It we have started with S' as focus Z'K'.

LATUS RECTUM

A line passing through the focus and parallel to the directrix of an ellipse is called the latus rectum of the ellipse.

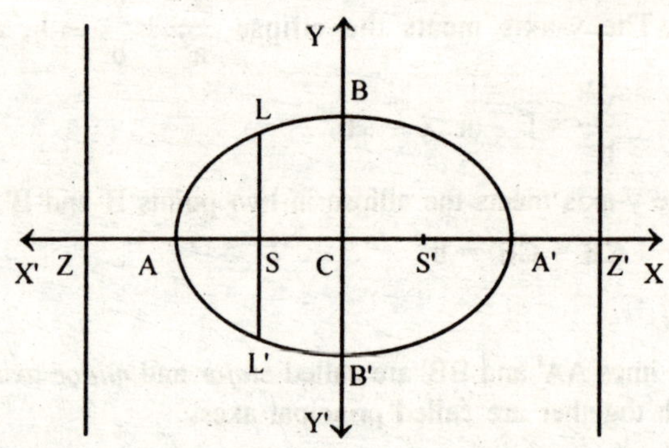

Let the line perpendicular at the focus meets the ellipse in L and L', the LSL' is called latus rectum. By symmetry. We have

$$LSL' = 2LS$$

If LS = y, then the co-ordinate of L is (–ae, y). As L lies on the ellipse, then we have

$$\frac{(-ae)^2}{a^2} + \frac{y^2}{b^2} = 1$$

so that $y^2 = b^2(1 - e^2) = \dfrac{b^4}{a^2}$ $\qquad \left(\because \; 1-e^2 = \dfrac{b^2}{a^2}\right)$

$y = \dfrac{b^2}{a}$, i.e., the semi-latus rectum of the ellipse is $\dfrac{b^2}{a}$.

FOCAL PROPERTY

Show that the sum of the focal distances of a point on the ellipse is constant and is equal to the major axis.

Proof:

Let P(x, y) be any point on the ellipse. Let S and S' be the two focii and ZM, Z'M be the two directrices. Join PS and PS'. Let PM, PM' and PL be perpendiculars from P an ZM Z'M' and an x-axis respectively.

we have SP = ePM = eZL = e(CL + CZ)

$= e(x + a/e) = ex + a.$ ...(i)

Also we have S'P = ePM' = e LZ' = e(CZ' − CL) = a − ex ....(ii)

Adding (i) and (ii), we get

SP + SP' = 2a = Major axis which is constant

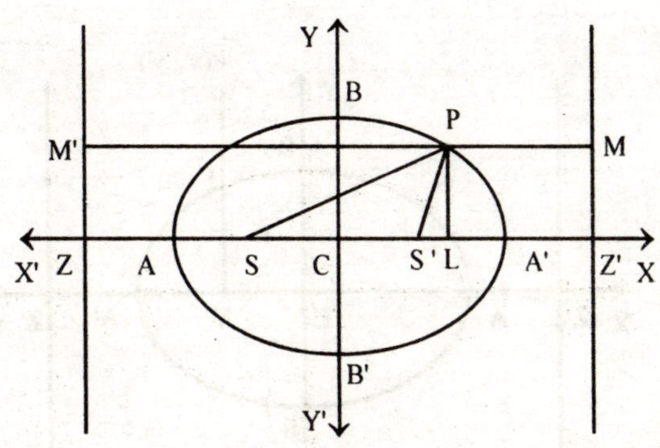

POSITION OF A POINT RELATIVE TO AN ELLIPSE

The point (x', y') is to lie outside, on or inside the ellipse $\frac{x^2}{a^2} + \frac{y^2}{b^2} = 1$ according as the quantity $\frac{x'^2}{a^2} + \frac{y'^2}{b^2} - 1$ is positive, zero, or negative.

Proof:

The point (x', y') lies on the ellipse $\frac{x^2}{a^2} + \frac{y^2}{b^2} - 1$

then $\frac{x'^2}{a^2} + \frac{y'^2}{b^2} - 1 = 0$

i.e. If the value is non zero then point is either inside the ellipse or outside the ellipse. Let P(x', y') lies out side the ellipse. Draw perpendicular PM meeting the ellipse on L.

The we have PM > LM i.e. $PM^2 - LM^2 > 0$ ...(i)

Since the point L is on the ellipse, its co-ordinates (x', LM) satisfy the equation of the ellipse, is given as i.e. $$\frac{x'^2}{a^2} + \frac{LM^2}{b^2} = 1$$

Now $$\frac{x'^2}{a^2} + \frac{y'^2}{a^2} - 1 > \frac{x'^2}{a^2} + \frac{LM^2}{b^2} - 1 \quad (\because y' > LM)$$

$$> 0 \left(\text{as } \frac{x'^2}{a^2} + \frac{LM^2}{b^2} - 1 = 0 \text{ due to } ...(ii) \right)$$

P(x', y')
Y
B
L
X' Z A C M A' Z' X
B'
Y'

Hence the condition for the point P to lie outside the ellipse

$$\frac{x^2}{a^2} + \frac{y^2}{a^2} = 1 \text{ is } \frac{x'^2}{a^2} + \frac{y'^2}{a^2} - 1 > 0$$

Similarly we can obtained the condition for the point to lie inside the ellipse

Example 1:

Find the equation of the ellipse (referred to its axes as the axes of x and y respectively) which passes through the point (–3, 1) and has the eccentricity $\sqrt{\frac{2}{5}}$.

Solution:

Let the required equation of the ellipse be

$$\frac{x^2}{a^2} + \frac{y^2}{b^2} = 1. \quad ...(i)$$

It passes through (–3, 1),

$$\therefore \qquad \frac{9}{a^2} + \frac{1}{b^2} = 1. \qquad \text{...(ii)}$$

Also, from the relation $b^2 = a^2(1 - e^2)$, we get

$$b^2 = a^2\left(1 - \frac{2}{5}\right)$$

$$\Rightarrow \qquad b^2 - \frac{3}{5}a^2. \qquad \text{...(iii)}$$

Solving equations (ii) and (iii), we get

$$a^2 = \frac{32}{3} \text{ and } b^2 = \frac{32}{5}$$

Hence the required equation becomes

$$3x^2 + 5y^2 = 32.$$

Example 2:

Show that $4x^2 + y^2 - 8x + 2y + 1 = 0$ represents an ellipse. Find its eccentricity, co-ordinates of its foci and the equation of its directrixes and latus rectum.

Solution:

The given equation can be written as

$$2^2(x- 1)^2 + (y +)^2 = 4' \qquad \text{...(i)}$$

$$\Rightarrow \qquad \frac{(x-1)^2}{1^2} + \frac{(x+1)^2}{2^2} = 1$$

Let $x - 1 = X$ and $y + 1 = Y$

∴ The equation becomes

$$\frac{X^2}{1^2} + \frac{Y^2}{2^2} = 1, \qquad \text{...(ii)}$$

which is the equation of the ellipse whose centre is $X = 0$, $Y = 0$ i.e. $x = 1$ and $y = -1$ and the minor and major axis he along

$$y + 1\ 0,\ x - 1 = 0$$

(ii) gives $a^2 = 4$, $b^2 = 1$

$$\therefore \qquad e^2 = 1 - \frac{b^2}{a^2} = 1 - \frac{1}{4} = \frac{3}{4}$$

i.e. $$e = \frac{\sqrt{3}}{2}$$

The foci are (0, – ae) and (0, ae) i.e. (0, √3), (0, √3) w.r.t. the equation (ii) or w.r.t. the point (1, – 1) as origin. Therefore the for with respect to origin (0, 0) are (1, – 1 + √3), (1, – 1 – √3).

The equation of the directrix referred to equation (ii) or w.r.t. to the point (1, – 1) as origin is

$$y = \pm \frac{a}{e} \text{ i.e. } y = -\frac{4}{\sqrt{3}} \text{ and } y = -\frac{4}{\sqrt{3}}.$$

∴ The equation of the directrix with reference to (0, 0) as origin is

$$y = \frac{4}{\sqrt{3}} - 1,\ y = -\frac{4}{\sqrt{3}} - 1$$

∴ The latus rectum $= \frac{2b^2}{a} = 1.$

Example 2:

A bar of given length moves with its extremities on the fixed straight lines at right angles. Prove that any point of the rod describes an ellipse.

Solution:

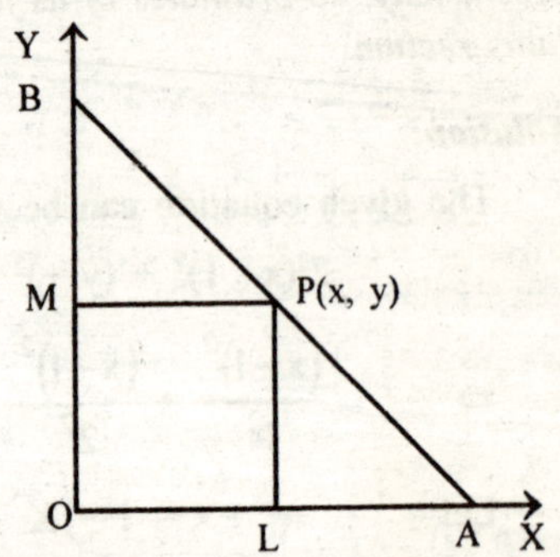

Consider two fixed straight lines as the co-ordinate axes. Let AB be the bar and P(x, y) any point on it. Let P divides AB internally in the ratios of m : n. From P draw PL ⊥ x-axis and PM ⊥ y-axis. If ∠OAB = θ, then

$$x = PM = \lambda n \cos\theta$$

$$y = PL = \lambda m \sin\theta$$

Eliminating θ from these equations we get

$$\frac{x^2}{\lambda^2 n^2} + \frac{y^2}{\lambda^2 m^2} = 1.$$

Hence the point P(x, y) describes an ellipse.

A LINE AND AN ELLIPSE

To find the point of intersection of the line y = mx + c with the ellipse

$$\frac{x^2}{a^2} + \frac{y^2}{b^2} = 1.$$

Proof:

The equation of the ellipse is

given as $$\frac{x^2}{a^2} + \frac{y^2}{b^2} = 1 \qquad ...(i)$$

and that of line is $$y = mx + c \qquad ...(ii)$$

The points of inter section of a line and an ellipse is obtained by solving above equation which gives

$$\frac{x^2}{a^2} + \frac{(mx+c)^2}{b^2} = 1$$

$$\Rightarrow \quad x^2(a^2m^2 + b^2) + 2a^2\,mcx + a^2(c^2 - b^2) = 0 \qquad ...(iii)$$

This equation being quadratic in x gives two values of x which are the abscissae of the points of intersection of (i) and (ii). Substituting these values of x in (iii) we get the corresponding values of y.

Thus a line meets an ellipse in two points real and distinct, coincident or imaginary according as the roots of equ. (iii) are real and distinct, coincident or imaginary.

To find the length of the intercept cut off by the straight line $y = mx + c$ form the ellipse $\frac{x^2}{a^2} + \frac{y^2}{b^2} = 1$.

Proof:

The equation of the line is given by

$$y = mx + c \qquad ...(i)$$

and that of the ellipse given by

$$\frac{x^2}{a^2} + \frac{y^2}{b^2} = 1. \qquad ...(ii)$$

Eliminating y from (i) and (ii), we have

$$\frac{x^2}{a^2} + \frac{(mx+c)^2}{b^2} = 1$$

$$\Rightarrow \quad x^2(a^2m^2 + b^2) + 2a^2\,mcx + a^2(c^2 - b^2) = 0 \qquad ...(iii)$$

The equation being a quadratic in x give two values of x say x_1 and x_2. Let the corresponding values of y be y_1, y_2.

$\therefore$ The points of intersection are (x_1, y_1) and (x_2, y_2). Equation (iii) give as

$$x_1 = x_2 = -\frac{2a^2mc}{a^2m^2+b^2}$$

and
$$x_1x_2 = \frac{a^2\left(c^2-b^2\right)}{a^2m^2+b^2}$$

Now $(x_1 - x_1)^2 = (x_2 - x_1)^2 - 4x_1x_2$

$$= \frac{2a^2m^2c^2}{\left(a^2m^2+b^2\right)} \frac{4a^2\left(c^2-b^2\right)}{a^2m^2+b^2}.$$

$$= \frac{4a^4m^2c^2\,4a^2\left(c^2-b^2\right)\left(a^2m^2+b^2\right)}{\left(a^2m^2+b^2\right)}$$

$$= \frac{4a^2}{\left(a^2m^2+b^2\right)^2}\;[a^2m^2c^2 - (a^2m^2c^2 + b^2c^2 - a^2m^2b^2 - b^4)]$$

$$\therefore (x_2 - x_1) = \frac{2ab}{a^2m^2+b^2}\sqrt{-b^2c^2+a^2m^2b^2+b^4}$$

$$= \frac{2ab}{a^2m^2+b^2}\sqrt{a^2m^2+b^2-c^2}$$

Again since (x_1, y_1) and (x_2, y_2) lie on (i)

$\Rightarrow$ $y_1 = mx_1 + c$

$y_2 = mx_2 + c$

Subtracting, we have

$$y_2 - y_1 = m(x_2 - x_1).$$

$$\therefore \text{Length of intercept} = \sqrt{(x_2-x_1)^2+(y_2-y_1)^2}$$

$$= \sqrt{(x_2-x_1)^2+m^2(x_2-x_1)^2}$$

$$= \sqrt{1+m^2}\,(x_2 - x_1)$$

$$= \sqrt{1+m^2}\left\{\frac{2ab}{a^2m^2+b^2}\sqrt{a^2m^2+b^2-c}\right\}$$

$$\frac{2ab}{a^2m^2+b^2}\sqrt{a^2m^2+b^2-c}\,\sqrt{1+m^2} \qquad \textbf{Ans.}$$

Cor. The line y = mx + c will be tangent to the ellipse if the intercept made by this line is zero i.e.

if $$\frac{2ab}{a^2m^2+b^2}\sqrt{a^2m^2+b^2-c}\,\sqrt{1+m^2} = 0$$

$\Rightarrow$if $c = \sqrt{a^2m^2+b^2}$.

This condition is also obtained independently in the following article.

CONDITION OF TANGENCY

To find the condition that the line y = mx + c may touch the ellipse $\frac{x^2}{a^2}+\frac{y^2}{b^2}=1$ *and hence find the equation of the tangent.*

Proof:

The equation of the line is given as

$$y = mx + c \qquad ...(i)$$

and the equation of the ellipse is

$$\frac{x^2}{a^2}+\frac{y^2}{b^2}=1. \qquad ...(ii)$$

Eliminating y from (i) and (ii), we have

$$\frac{x^2}{a^2}+\frac{(mx+c)^2}{b^2}=1$$

$$\Rightarrow \quad x^2(a^2m^2+b^2)+2a^2\,mcx+a^2(c^2-b^2)=0 \qquad ...(iii)$$

If the line (i) touches the ellipse (ii), the equation (iii) has equal roots and the condition for which is given by

$$4a^2m^2c^2 = 4(a^2m^2+b).\ a^2\,(c^2-b^2)$$

$$\Rightarrow \quad c^2 = a^2m^2+b^2$$

$$\Rightarrow \quad c = \pm\sqrt{a^2m^2+b^2},$$

which is the required condition.

Substituting the value of c in (ii), the equation of the tangent is

$$y = mx \pm \sqrt{a^2m^2 + b^2}\,.$$

This is also called the equation of tangent in slope form.

EQUATION OF TANGENT AND NORMAL

Equation of Tangent : *To find the equation of a tangent at a point (x_1, y_1) to the ellipse* $\frac{x^2}{x^2} + \frac{y^2}{b^2} = 1$.

Proof:

The equation of the ellipse is given as

$$\frac{x^2}{a^2} + \frac{y^2}{b^2} = 1 \qquad \text{...(i)}$$

Let $Q(x_2, y_2)$ be another point on the ellipse in the neighbourhood of the point $P(x_1, y_1)$

The equation of the chord joining the two points (x_1, y_1) and (x_2, y_2) is given as

$$y - y_1 = \frac{y_2 - y_1}{x_2 - x_1}\,(x - x_1) \qquad \text{...(ii)}$$

Since (x_1, y_1) and (x_2, y_2) lie on the ellipse (i), then we have

$$\therefore \quad \frac{x_1^2}{a^2} + \frac{y_1^2}{b^2} = 1 \qquad \text{...(iii)}$$

and

$$\frac{x_2^2}{a^2} + \frac{y_2^2}{b^2} = 1. \qquad \text{...(iv)}$$

Substracting (iii) from (iv), we obtain

$$\frac{x_2^2 - x_1^2}{a^2} + \frac{y_2^2 - y_1^2}{b^2} = 0$$

$$\Rightarrow \quad \frac{(x_2 - x_1)(x_2 - x_1)}{a^2} + \frac{(y_2 - y_1)(y_2 - y_1)}{b^2} = 0$$

$$\Rightarrow \quad \frac{y_2 - y_1}{x_2 - x_1} = -\frac{b_2}{a^2}\left(\frac{x_2 + x_1}{y_2 + y_1}\right) \qquad \text{...(iv)}$$

From (ii) and (v), we obtain

$$y - y_1 = -\frac{b_2}{a^2}\frac{x_2 + x_1}{y_2 + y_1}(x - x_1) \qquad \text{...(vi)}$$

The chord becomes tangent at (x_1, y_1) when $x_2 \to x_1$ and $y_2 \to y_1$

$\therefore$ From (vi), the equation of the tangent is given as

$$y - y_1 = -\frac{b_2}{a^2}\frac{2x_1}{2y_1}$$

$$\frac{yy_1}{b^2} - \frac{y_1^2}{b^2} = \frac{xx_1}{x^2} + \frac{x_1^2}{a^2}$$

$$\Rightarrow \quad \frac{xx_1}{a^2} + \frac{yy_1}{b^2} = \frac{x_1^2}{a^2} + \frac{y_1^2}{a^2} = 1$$

$$\Rightarrow \quad \frac{xx_1}{a^2} + \frac{yy_1}{b^2} = 1$$

which is the required equation of tangent

EQUATION OF NORMAL

To find the equation of the normal at the point (x_1, y_1) to the ellipse

$$\frac{x^2}{a^2} + \frac{y^2}{b^2} = 1.$$

Proof:

The equation the ellipse is given as

$$\frac{x^2}{a^2} + \frac{y^2}{b^2} = 1. \qquad \text{...(i)}$$

The equation of the tangent at (x_1, y_1) to the ellipse (i) is given by

$$\frac{xx_1}{a^2} + \frac{yy_1}{b^2} = 1.$$

Then slope of the tangent $= \dfrac{\frac{x_1}{a^2}}{\frac{y_1}{b^2}} = -\dfrac{b^2}{a^2}\dfrac{x_1}{y_1}.$

$\therefore$ Slope of the normal $= \dfrac{a^2}{b^2}\dfrac{y_1}{x_1}.$

$\therefore$ Equation of the normal at the point (x_1, y_1) is

$$y - y_1 = \frac{a^2 y_1}{b^2 x_1}(x - x_1)$$

$$\Rightarrow \quad \frac{yb^2}{y_1} - b^2 = \frac{a^2 x}{x_1} - a^2$$

or $$\frac{a^2 x}{x_1} - \frac{b^2 y}{y_1} = a^2 - b^2,$$

which is the required equation of Normal

Example 1:

Prove that the line $x \cos \alpha + y \sin \alpha = p$ is a tangent to the ellipse

$$\frac{x^2}{a^2} + \frac{y^2}{b^2} = 1$$

if $p^2 = a^2 \cos^2 \alpha + b^2 \sin^2 \alpha$.

Solution:

Any tangent to the ellipse

$$\frac{x^2}{a^2} + \frac{y^2}{b^2} = 1 \text{ is given as } \quad y = mx + \sqrt{a^2m^2 + b^2}$$

Comparing it with the equation

$$x \cos \alpha + y \sin \alpha = p$$

or $$y = -\frac{\cos\alpha}{\sin\alpha} x + \frac{p}{\sin x}$$

we get, $$m = -\frac{\cos\alpha}{\sin\alpha}, \frac{p}{\sin\alpha} = \sqrt{a^2m^2 + b^2}.$$

Eliminating m, we have

$$\frac{p}{\sin\alpha} = a^2 \frac{\cos\alpha}{\sin\alpha} + b^2$$

$$\Rightarrow \quad p^2 = a^2 \cos^2 \alpha + b^2 \sin^2 \alpha.$$

Example 2:

If the normal at the end of a latus rectum of an ellipse passes through one extremity of a minor axes, show that the eccentricity of the curve is given by the equation $e^4 + e^2 - 1 = 0$.

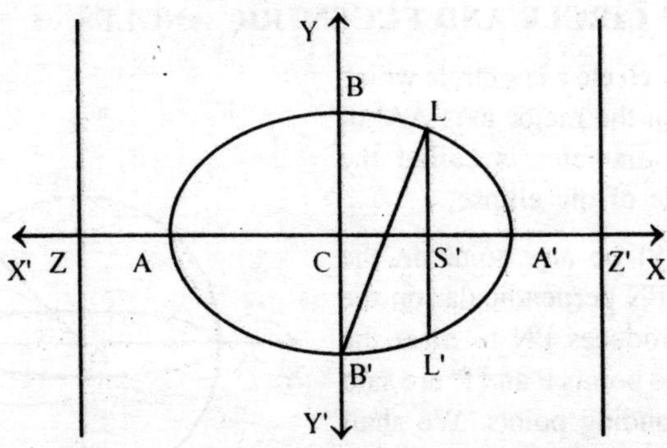

Solution:

Let the equation of the ellipse be

$$\frac{x^2}{a^2} + \frac{y^2}{b^2} = 1. \quad \text{...(i)}$$

The co-ordinate of the end of the latus rectum in the quadrant BCA' are

$$\left(ae, \frac{b^2}{a}\right).$$

Now the equation of the normal at

$$\left(ae, \frac{b^2}{a}\right) \text{ is}$$

$$\frac{a^2x}{ae} + \frac{b^2y}{b^2/a} = a^2 - b^2.$$

$$\Rightarrow \quad \frac{ax}{e} - ay = a^2 - b^2.$$

Now if this passes through B' (0, – b), then we have

$$0 + ab + a^2 - b^2$$

$$\Rightarrow \quad a^2b^2 = (a^2 - b^2)^2.$$

Substituting $b^2 = a^2(1 - e^2)$, we have

$$a^2 \times a^2(1 - e^2) = [a^2 - a^2(1 - e^2)]^2$$

$$\Rightarrow \quad 1 - e^2 = e^4$$

$$\Rightarrow \quad e^4 + e^2 - 1 = 0$$

AUXILIARY CIRCLE AND ECCENTRIC ANGLES

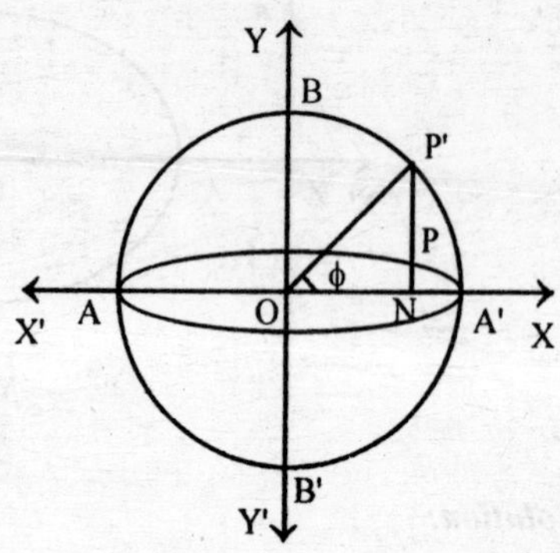

Auxiliary circle : The circle which is described on the major axis AA' of an ellipse as diameter is called the auxillary circle of the ellipse.

Let P(x, y) be any point on the ellipse. Draw PN perpendicular on the major axis. Produces PN to meet the circle in P'. The points P and P' are said to be corresponding points. We shall now prove that the ordinates of any point P on the ellipse and its corresponding point P' on the auxiliary circle are in constant ratio.

The equation of the ellipse is given as

$$\frac{x^2}{a^2} + \frac{y^2}{b^2} = 1 \qquad \text{...(i)}$$

and the equation of circle is given as

$$x^2 + y^2 = a^2. \qquad \text{...(ii)}$$

The co-ordinates of P and P' are (x, PN) and (x, PN)

Now P lies on ellipse then we have

$$\therefore \quad \frac{x^2}{a^2} + \frac{PN^2}{b^2} = 1 \qquad \text{...(iii)}$$

Also P' lies on circle then we have

$$\therefore \quad \frac{x^2}{a^2} + \frac{P'N}{a^2} = 1 \qquad \text{...(iv)}$$

Now (iii) and (iv) gives

$$\frac{PN^2}{b^2} + \frac{P'N^2}{a^2}$$

$$\Rightarrow \quad PN : P'N = b : a$$

Hence the ordinates of any point on the ellipse is b/a times the ordinate of corresponding point on the auxiliary circle.

ECCENTRIC ANGLE

The eccentric angle of a point P on the ellipse is defined as the angle between the major axis and the line joining the centre of the ellipse to the corresponding point of P i.e. P' on the auxiliary circle i.e. ∠P'CN is the eccentric angle of P. It is generally denoted by ϕ.

In the ΔP'CN

$$CN = P'C \cos\phi = a \cos\phi = \text{abscissa of } P.$$

Putting $x = a \cos\phi$ in the equation of the ellipse, we get

$$\frac{a^2 \cos^2\phi}{a^2} + \frac{y^2}{b^2} = 1$$

$$\Rightarrow \quad y^2 = b^2(1 - \cos^2\phi)$$

$$= b^2 \sin^2\phi$$

$$y = \pm\, b \sin\phi$$

∴ The co-ordinates of P are $(a \cos\phi, b \sin\phi)$. These co-ordinates are also called parametric co-ordinates of P on the ellipse and $x = a\cos\phi$, $y = b\sin\phi$ is the parametric equation of the ellipse.

EQUATION OF THE CHORD OF THE ELLIPSE JOINING THE POINTS WHOSE ECCENTRIC ANGLES ARE Q AND F

The equation of the chord of the ellipse joining two points $(a \cos\phi, b \sin\phi)$ and $(a \cos\phi, b \sin\phi)$ on the ellipse $\frac{x^2}{a^2} + \frac{y^2}{b^2} = 1$ is

$$y - b\sin\phi = \frac{b(\sin\phi - \sin\theta)}{a(\cos\phi - \cos\theta)}\,(x - a\cos\phi)$$

$$\frac{y - b\sin\theta}{b}(\cos\phi - \cos\theta) = (\sin\phi - \sin\theta)\frac{(x - a\cos\theta)}{a}$$

$$\Rightarrow \quad \frac{y - b\sin\theta}{b}\sin\frac{\theta+\phi}{2} = -\cos\frac{\theta+\phi}{2}\left(\frac{x - a\cos\theta}{a}\right)$$

$$\Rightarrow \quad \frac{y}{b}\sin\frac{\theta+\phi}{2} - \sin\theta\sin\frac{\theta+\phi}{2}\;\frac{-x}{a}\cos\frac{\theta+\phi}{2} + \cos\frac{\theta+\phi}{2}\cos\theta$$

$$\Rightarrow \quad \frac{x}{a}\cos\frac{\theta+\phi}{2} + \frac{y}{b}\sin\frac{\theta+\phi}{2} = \cos\left(\theta - \frac{\theta+\phi}{2}\right)$$

$$\Rightarrow \frac{x}{a}\cos\frac{\theta+\phi}{2}+\frac{y}{b}\sin\frac{\theta+\phi}{2}=\cos\frac{\theta-\phi}{2} \quad \text{...(i)}$$

which is the required equation.

Cor. Equation of Tangent and Normal at the Point θ

The chord (i) will be tangent to the ellipse at the point θ if θ = ϕ. Putting θ = ϕ in (i), we get

$$\frac{x}{a}\cos\theta+\frac{y}{b}\sin\theta=1$$

which is the equation of the tangent to the ellipse at the point θ.

Similarly, the equation of the normal at the point θ is given as

$$ax\sec\theta - by\,\mathrm{cosec}\,\theta = (a^2-b^2).$$

CONORMAL POINTS

Show that four normals can be drawn to an ellipse from any point and the sum of the eccentric angles of the conormal points is equal to an odd multiple of the right angles.

Let the equation of the ellipse be given as

$$\frac{x^2}{a^2}+\frac{y^2}{b^2}=1. \quad \text{...(i)}$$

The normal at any point q i.e.. (a cos θ, b sin θ) on (i) is given by

$$\frac{ax}{\cos\theta}-\frac{by}{\sin\theta}=a^2-b^2.$$

Let (h, k) be a fixed point through which the normal passes then we have

$$\frac{ah}{\cos\theta}-\frac{bk}{\sin\theta}=a^2-b^2. \quad \text{...(ii)}$$

Using the transformation we have

$$\sin\theta=\frac{2t}{1+t^2};\ \cos\theta=\frac{1-t^2}{1+t^2}$$

where $t=\tan\frac{\theta}{2}$; the equation then equation (ii) becomes

$$ah\left[\frac{1+t^2}{1+t^2}\right]-bk\left[\frac{1-t^2}{1+t^2}\right]=a^2-b^2$$

$$\Rightarrow \quad 2ath\,(1 + t^2) - bk(1 + t^2) = 2t(1 - t^2)\,(a^2 + b^2)$$

$$\Rightarrow \quad bkt^4 + 2(ah + a^2 - b^2)\,t^3 + 2(ah - a^2 + b^2)\,t - bk = 2. \qquad \text{...(iii)}$$

This equation is biquadratic in 't' and therefore it has four roots, real or imaginary.

Let the four roots be t_1, t_2, t_3, t_4 and the corresponding values of θ_1, θ_2, θ_3, θ_4. Now corresponding to each value of θ, we get one point on the ellipse such that the normal at this point passes through the fixed point (h, k). This proves that in general four normals can be drawn from a point to an ellipse.

The roots of (iii) are tan

$$\frac{\theta_1}{2}, \tan\frac{\theta_2}{2}, \tan\frac{\theta_3}{2}, \tan\frac{\theta_4}{2}$$

$$\therefore \quad S_1 = \Sigma \tan\frac{\theta_1}{2} = -\frac{2\left(ah + a^2 - b^2\right)}{bk}$$

$$S_2 = \Sigma \tan\frac{\theta_1}{2}, \tan\frac{\theta_2}{2} = 0$$

$$S_3 = \Sigma \tan\frac{\theta_1}{2}\tan\frac{\theta_2}{2}\tan\frac{\theta_3}{2} = -\frac{2\left(ah - a^2 + b^2\right)}{bk}$$

$$S_3 = \Sigma \tan\frac{\theta_1}{2}\tan\frac{\theta_2}{2}\tan\frac{\theta_3}{2}\tan\frac{\theta_4}{2} = -1$$

Now we have

$$\tan\left(\frac{\theta_1}{2} + \frac{\theta_2}{2} + \frac{\theta_3}{2} + \frac{\theta_4}{2}\right) = \frac{S_1 - S_3}{1 - S_2 + S_4} = \frac{S_1 - S_3}{1 - 0 + 1} = \infty = \tan\frac{\pi}{2}$$

$$\Rightarrow \quad \frac{\theta_1}{2} + \frac{\theta_2}{2} + \frac{\theta_3}{2} + \frac{\theta_4}{2} = n\pi + \frac{\pi}{2}$$

$$\text{or} \quad \theta_1 + \theta_2 + \theta_3 + \theta_4 = 2n\pi = \pi = (2n + 1)\,\pi,$$

which is an odd multiple of π.

Example 1:

If α, β,γ be the eccentric angles of three point on an ellipse, the normals at which are concurrent, show that

$$\sin(\alpha + \beta) + \sin(\beta + \gamma) + \sin(\gamma + \alpha) = 0.$$

Solution:

Let the equation of the ellipse be $\frac{x^2}{a^2} + \frac{y^2}{b^2} = 1$.

The normal at the point (α cos α, b sin α) is

$$ax \sec \alpha - by \operatorname{cosec} \alpha = a^2 - b^2.$$

$$\Rightarrow \quad ax \sin \alpha - by \cos \alpha = (a^2 - b^2) \sin \alpha \cos \alpha$$

$$\Rightarrow \quad ax \sin \alpha - by \cos \alpha + \frac{1}{2}(b^2 - a^2) \sin 2\alpha = 0 \qquad \text{...(i)}$$

Similarly, the other two normals are

$$a\alpha \sin \beta - by \cos \beta + \frac{1}{2}(b^2 - a^2) \sin 2\beta = 0. \qquad \text{...(ii)}$$

$$a\alpha \sin \gamma - by \cos \gamma + \frac{1}{2}(b^2 - a^2) \sin 2\gamma = 0. \qquad \text{...(iii)}$$

The line are concurrent

$$\therefore \begin{vmatrix} \sin\alpha & \cos\alpha & \sin 2\alpha \\ \sin\beta & \cos\beta & \sin 2\beta \\ \sin\gamma & \cos\gamma & \sin 2\gamma \end{vmatrix} = 0$$

$$\Rightarrow \sin 2\alpha(\sin \beta \cos \gamma - \cos \beta \sin \gamma) - \sin 2\beta(\sin \alpha \cos \gamma - \cos \alpha \sin\gamma) + \sin 2\gamma(\sin \alpha \cos \beta - \cos \alpha \sin \beta) = 0.$$

$$\Rightarrow \sin 2\alpha \sin \beta \cos \gamma = \sin 2\beta \sin \gamma \cos \alpha + \sin 2\gamma \sin \alpha \cos \beta$$

$$= \sin 2\alpha \cos \beta \sin \gamma + \sin 2\beta \cos \gamma \sin \alpha + \sin 2\gamma \cos \alpha \sin \beta.$$

Adding $\sin^2\alpha \cos^2\gamma + \sin^2\beta \cos^2\alpha + \sin^2\gamma \cos^2\beta$ to L.H.S., and the equal expression $\cos^2\alpha \sin^2\gamma + \cos^2\beta \sin^2\alpha + \cos^2\gamma \sin^2\beta$ to L.H.S., we get

$$(\sin \alpha \cos \gamma + \sin \beta \cos \alpha + \sin \gamma \cos \beta)^2$$

$$= (\cos \alpha \sin \gamma + \cos \gamma \sin \alpha + \cos \gamma \sin \beta)^2$$

$$\Rightarrow (\sin \alpha \cos \gamma + \sin \beta \cos \alpha + \sin \gamma \cos \beta)^2$$

$$= (\cos \alpha \sin \gamma + \cos \beta \sin \alpha + \cos \gamma \sin \beta)^2 = 0$$

$$\Rightarrow (\Sigma \sin \alpha \cos \gamma + \Sigma \cos \alpha \cos \gamma)$$

$$(\Sigma \sin \alpha \cos \gamma - \Sigma \cos \alpha \sin \gamma) = 0$$

$$\Rightarrow \{\sin (\alpha + \gamma) + \sin (\beta + \alpha) + \sin (\gamma + \beta)\}$$

$$\{\sin (\alpha - \gamma) + \sin (\beta - \alpha) + \sin (\gamma - \beta)\} = 0$$

But $\sin (\alpha - \gamma) + \sin (\beta - \alpha) + \sin (\gamma - \beta)$

$$= 4 \sin \frac{1}{2} (\alpha - \gamma) \sin \frac{1}{2} (\beta - \alpha) \sin \frac{1}{2} (\gamma - \beta)$$

$\neq 0$ as $\alpha \neq \beta \neq \gamma$ (using identities)

$\therefore \sin (\alpha + \gamma) + \sin (\beta + \alpha) + \sin (\gamma + \beta) = 0.$

PROPERTIES OF ELLIPSE

(i) The tangent and the normal at any point of an ellipse bisect the angle between the focal radii to that point.

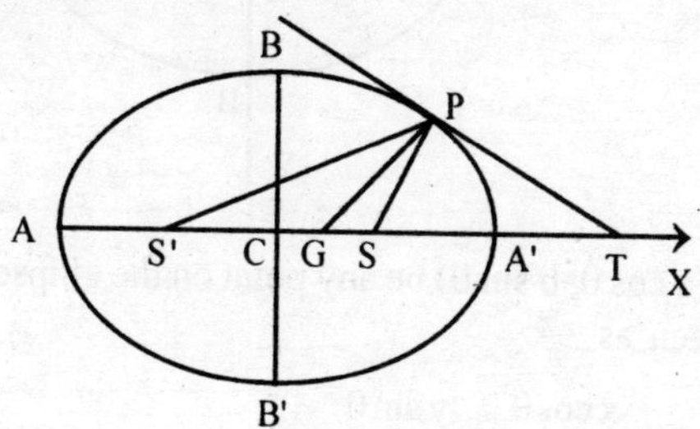

Let $P(x_1, y_1)$ be a point on the ellipse $\frac{x^2}{a^2} + \frac{y^2}{b^2} = 1$ and let the tangent and normal at P meet the major axis in T and G respectively.

Equation to the normal at P is given as

$$\frac{x - x_1}{x_1 / a^2} = \frac{y - y_1}{y_1 / b^2}$$

It meets the x-axis i.e. $y = 0$ then we have

$$x = x_1 \left(\frac{b^2}{a^2} - 1 \right) = e^2 x_1.$$

This gives $OG = e^2 x_1$.

$\therefore$ $S'G = ae + e^2 x_1 = e(a + ex_1) = eS'P$

$SG = ae + e^2 x_1 = e(a - ex_1) = eSP.$

$\therefore$ $\frac{S'G}{SG} = \frac{S'P}{SP}.$

Hence, PG is the bisector of the angle S' PS. Since the tangent is perpendicular to the normal, PT bisects the angle S' PS externally.

(ii) If SW and S' W be perpendicular from the foci upon the tangent at any point P of the ellipse then W and W' lie on the auxiliary circle and SW. S' W' = b^2.

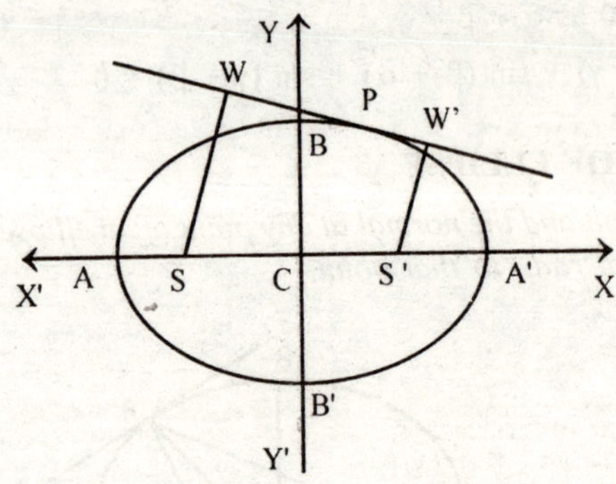

Let P(a cos θ, b sin θ) be any point on the ellipse. The equation of tangent at P is given as

$$\frac{x\cos\theta}{a}+\frac{y\sin\theta}{b}=1 \qquad \text{...(i)}$$

Equation of S' W' passing through (ae, 0) and perpendicular to (i) is given by solving equations (i) and (ii). Squaring and adding, we get

$$(x^2+y^2)\left(\frac{\cos^2\theta}{a^2}+\frac{\sin^2\theta}{b^2}\right)=1+\frac{e^2a^2}{b^2}\sin^2\theta$$

$$=\frac{b^2+(a^2-b^2)\sin^2\theta}{b^2}=\frac{b^2\cos^2\theta+a^2\sin^2\theta}{b^2}$$

$\therefore \quad x^2+y^2=a^2.$

It shows that W', the point of intersection of (i) and (ii) lies on the auxiliary circle. Similarly, writing the equation SW and solving the equation with (i), it can be shown that W also lies on the auxiliary circle.

We have S' W' = the perpendicular distance from S(ae, 0) on the line (i).

$$=\frac{1-e\cos\theta}{\sqrt{\frac{\cos^2\theta}{a^2}+\frac{\sin^2\theta}{b^2}}}=\frac{ab(1-e\cos\theta)}{\sqrt{(a^2\sin^2 a+b^2\cos^2\theta)}}$$

Similarly, $\quad S'W=\dfrac{ab(1-e\cos\theta)}{\sqrt{(a^2\sin^2\theta+b_2\cos^2\theta)}}$

$$\therefore \qquad SW.\ S'W = \frac{a^2b^2\left(1-e^2\cos^2\theta\right)}{\left(a^2\sin^2\theta+b^2\cos^2\theta\right)}$$

$$= \frac{a^2b^2\left(1-e^2\cos^2\theta\right)}{a^2\left(1-e^2\cos^2\theta\right)} = b^2$$

Hence the result

(iii) If SW and S' W' be two perpendiculars from the foci upon the tangent at any point P of an ellipse then CW is parallel to S'P and CW' is parallel to SP.

Let P(h, k) be any point on the ellipse

$$\frac{x^2}{a^2}+\frac{y^2}{b^2}=1.$$

The equation of tangent at P is given as

$$\frac{xh}{a^2}+\frac{yk}{b^2}=1.$$

It meets x-axis at $y = 0$ i.e. at $x = \frac{a^2}{h}$

$\therefore CT = \frac{a^2}{h}$. Then we have

$$S'T = S'C + CT = ae + \frac{a^2}{h} = \frac{aeh+a^2}{h}$$

$$= \frac{a(eh+a)}{h} = \frac{aS'P}{h}$$

$$\therefore \qquad \frac{CT}{S'T} = \frac{a}{S'P} = \frac{CW}{S'P}$$

(we have proved that W lies on the auxiliary circle)

$x^2 + y^2 = a^2$ i.e. CW = radius = a

$\Rightarrow$ CW is parallel to S' P. Hence the result

(iv) If the normal at any point P of the ellipse meets the major and minor axes in G and g and if CF be perpendicular from the centre upon the normal then

$$PF\ .\ PG = b^2 \text{ and } PF\ .\ Pg = a^2$$

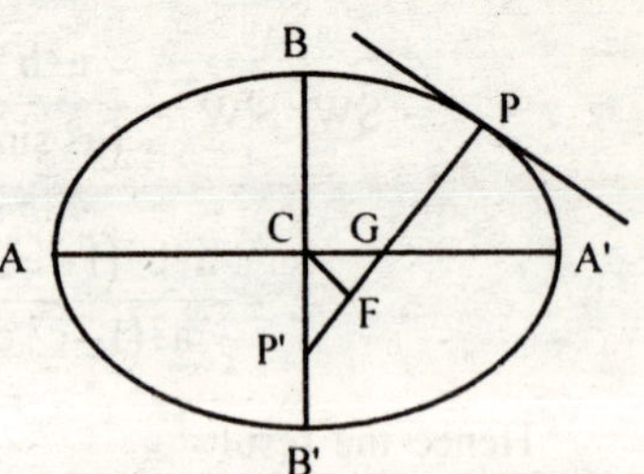

Let the equation of the ellipse be

$\frac{x^2}{a^2} + \frac{y^2}{b^2} = 1$ and $P(x_1, y_1)$ be any point on it. The equation of tangent at P is given as

$$\frac{xx_1}{a^2} + \frac{yy_1}{b^2} = 1 \qquad ...(i)$$

Let CQ be perpendicular to the tangent at P the we have

PF = CQ = perpendicular from (0, 0) on (i)

$$= \frac{1}{\sqrt{\left(\frac{x_1^{3}}{a} + \frac{y_1^{2}}{b^4}\right)}}$$

Co-ordinates of G are $(e^2x_1, 0)$

$$\therefore\ PG = \sqrt{(x_1 - e^2x_1)^2 + y_1^2} = \sqrt{x_1^2(1-e^2)^2 + y_1^2}$$

$$= \sqrt{x_1^2 \frac{b^4}{a^2} + y_1^2} \qquad \left(\because (1-e^2) = \frac{b^2}{a^2}\right)$$

$$= b^2\sqrt{\left(\frac{x_1^2}{a^4} + \frac{y_1^2}{b^4}\right)}$$

$\therefore$ PF . PG = b^2

Similarly, we can obtain the co-ordinates of g and then we have

PF . Pg = a^2

Hence the result.

DIRECTOR CIRCLE

Definition: *It is the locus of the point of intersection of two perpendicular tangents to an ellipse.*

Or

A circle whose centres is the centres of the ellipse and whose radius is the length of the line joining the ends of the major and mirror axis. This circle is called the Director circle.

To Find the Equation of Director Circle

We have that through any point $P(x_1, y_1)$, two tangents are given by the equation

$$m^2(x_1^2 - a^2) - 2mx_1y_1 + (y_1^2 - b^2) = 0.$$

The two tangents can be at right angle if the product of their slopes is -1 i.e.

$$m_1 \times m_2 = -1$$

$$\Rightarrow \quad \frac{y_1^2 - b^2}{x_1^2 - a^2} = -1 \text{ or } (x_1^2 - a^2) + (y_1^2 - a^2) = 0$$

$$\Rightarrow \quad x_1^2 + y_1^2 = a^2 + b^2.$$

$\therefore$ the locus of the point P is

$$x^2 + y^2 = a^2 + b^2, \text{ which is a circle.}$$

PAIR OF TANGENTS FROM A POINT

Let $h(x_1, y_1)$ be any point. From h draw a line hP_1P_2. Let $P(x, y)$ be any point. Let P_1 be the point which divides the join of hP in the ratio $\lambda : 1$. The co-ordinates of the point P_1 are $\left(\frac{\lambda x + x_1}{\lambda + 1}, \frac{\lambda y + y_1}{\lambda + 1}\right)$

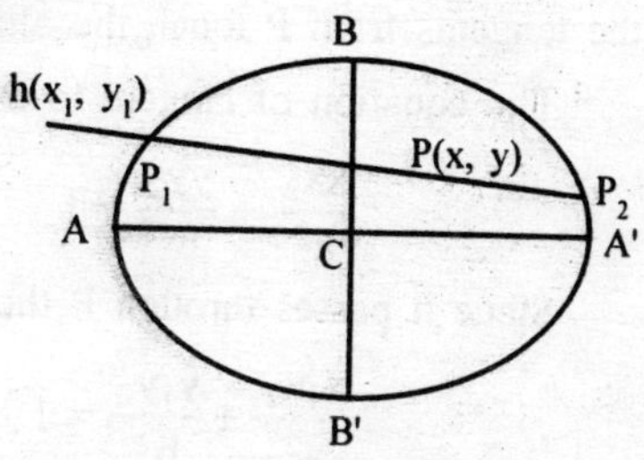

Since P_1 lies on the ellipse, so we obtain

$$\frac{\left(\frac{\lambda x + x_1}{\lambda + 1}\right)^2}{a^2} + \frac{\left(\frac{\lambda y + y_1}{\lambda + 1}\right)^2}{b^2} = 1$$

$$\Rightarrow \quad \frac{1}{a^2}(\lambda x + x_1)^2 + \frac{1}{b^2}(\lambda y + y_1)^2 = (\lambda + 1)^2$$

$$\Rightarrow \quad \lambda^2\left[\frac{x^2}{a^2} + \frac{y^2}{b_2} - 1\right] + 2\lambda\left[\frac{xx_1}{a^2} + \frac{yy_1}{b_2} - 1\right]$$

$$+\left(\frac{x_1^2}{a^2} + \frac{y_1^2}{b^2} - 1\right) = 0. \quad \text{...(i)}$$

The above equation is quadrate in λ, which gives two values of λ, one corresponds to the point P_1 and the other to the point P_2. The line will become tangent to the ellipse if both the values of λ given by (i) are same i.e., if

$$\left(\frac{x^2}{a^2}+\frac{y^2}{b^2}-1\right)\left(\frac{x_1^2}{a^2}+\frac{y_1^2}{b^2}-1\right)=\left(\frac{xx_1}{a^2}+\frac{yy_1}{b^2}-1\right)^2$$

which is the required equation of the pair tangents.

CHORD OF CONTACT

Definition: Let PQ and PR be two tangents from P the ellipse at the point Q and R. The chord joining Q and R is called the chord of contact of the point P.

TO FIND THE EQUATION OF THE CHORD OF CONTACT

Let the ellipse be $\frac{x^2}{a^2}+\frac{y^2}{b^2}=1$ and $P(x_1, y_1)$ be an external point. Let the tangents from P touch the ellipse at the point $P(x_2, y_2)$ and $R(x_3, y_3)$.

The equation of tangent to the ellipse at $Q(x_2, y_2)$ is given by

$$\frac{xx_2}{a^2}+\frac{yy_2}{b^2}=1. \qquad \text{...(i)}$$

Since it passes through P then we have

$$\therefore \quad \frac{x_1x_2}{a^2}+\frac{y_1y_2}{b^2}=1. \qquad \text{...(ii)}$$

Also, the equation of tangent at $R(x_3, y_3)$ passes through P the we have

$$\frac{x_1x_3}{a^2}+\frac{y_1y_3}{b^2}=1. \qquad \text{...(iii)}$$

Equation (ii) and (iii) show that the points (x_2, y_2) and (x_3, y_3) lie on the straight line then we have

$$\frac{xx_1}{a^2}+\frac{yy_1}{b^2}=1,$$

which is the required equation of the chord of contact of the tangents drawn from the point (x_1, y_1) to the ellipse

$$\frac{x^2}{a^2}+\frac{y^2}{b^2}=1.$$

POLE AND POLAR

The polar of a point P with respect to an ellipse is the locus of the point of intersection of tangents drawn at the extremities of the chords through P. The point P is called the polar.

TO FIND THE POLAR OF THE POINT $P(x_1, y_1)$ WITH RESPECT TO AN ELLIPSE

Let the equation of the ellipse be $\frac{x^2}{a^2} + \frac{y^2}{b^2} = 1$ and let $P(x_1, y_1)$ be the given point. Through P draw a chord QR. Let S be the point of intersection of the tangents at Q and R, then the locus of the point S is called the polar of P. Let the co-ordinates of S be (h, k). Now QR is the chord of contact of the tangents drawn from S to the ellipse, therefore the equation of QR, the chord of contact is given by

$$\frac{xh}{a^2} + \frac{yk}{b^2} = 1.$$

Since it passes through the point $P(x_1, y_1)$ then we have

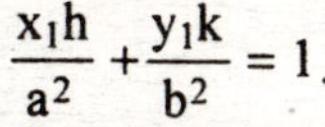

$$\frac{x_1h}{a^2} + \frac{y_1k}{b^2} = 1.$$

Hence the locus of S(h, k) is given by

$$\frac{xx_1}{a^2} + \frac{yy_1}{b^2} = 1,$$

which is the polar of the point $P(x_1, y_1)$ with respect to the ellipse.

POLE OF A LINE

To find the pole of the line Ax + By + C = 0 with respect to the ellipse

$$\frac{x^2}{a^2} + \frac{y^2}{b^2} = 1$$

Let $P(x_1, y_1)$ be the pole of the given line

$$Ax + By + C = 0. \qquad \text{...(i)}$$

with respect to the ellipse

$$\frac{x^2}{a^2}+\frac{y^2}{b^2}=1. \qquad ...(ii)$$

The equation of polar of $P(x_1, y_1)$ w.r.t. (ii) is given as

$$\frac{xx_1}{a^2}+\frac{yy_1}{b^2}=1. \qquad ...(iii)$$

Now the equations (i) and (iii) represent the same line,

$\therefore$ comparing the coefficients, we have

$$\frac{\frac{x_1}{a^2}}{A}=\frac{\frac{y_1}{b^2}}{B}=-\frac{1}{C}$$

$$\Rightarrow \qquad x_1=-\frac{a^2A}{C},\ y_1=-\frac{b^2B}{C}.$$

Hence the required pole is $\left(-\frac{a^2A}{C}, -\frac{b^2B}{C}\right)$.

CONJUGATE POINTS

To show that if the polar of a point P with respect to an ellipse passes through Q, then the polar of Q, also passes through P.

Let the equation of the ellipse be given as

$$\frac{x^2}{a^2}+\frac{y^2}{b^2}=1 \qquad ...(i)$$

and the co-ordinates of P Q are (x_1, y_1) and (x_2, y_2) respectively.

The polar of point P with respect to ellipse (i) is given by

$$\frac{xx_1}{a^2}+\frac{yy_1}{b^2}=1.$$

It passes through Q then we have

$$\frac{x_1x_2}{a^2}+\frac{y_1y_2}{b^2}=1. \qquad ...(ii)$$

The polar of point Q with respect to the ellipse is given as

$$\frac{xx_1}{a^2}+\frac{yy_1}{b^2}=1.$$

It will pass through P if

$$\frac{x_1x_2}{a^2}+\frac{y_1y_2}{b^2}=1 \text{, which holds because of (ii).}$$

Hence the polar of Q passes through P. The points P and Q are called conjugate points.

CONJUGATE LINES

To the pole of the line $l_1x + m_1y + n_1 = 0$ with respect to an ellipse lies on the line $l_2x + m_2y + n_2 = 0$, then pole of line $l_2x + m_2y + n_2 = 0$ with respect to the same ellipse lies on the line $l_1x + m_1y + n_1 = 0$.

Proof:

Let the pole of the line $l_1x + m_1y + n_1 = 0$ with respect to the ellipse

$\frac{x^2}{a^2}+\frac{y^2}{b^2}=1$ be $P(x_1, + y_1)$, then we have

$$x_1 = -\frac{a^2l_1}{n_1}, \quad y = -\frac{b^2m_1}{n_1}.$$

The point P lies on the line $l_2x + m_2y + n_2 = 0$

$$\therefore \quad \left(-\frac{a^2l_1}{n_1}\right)l_2+\left(-\frac{b^2m_1}{n_1}\right)m_2+n_2=0$$

$$\Rightarrow \quad a^2l_1l_2 + b^2m_1m_2 = n_1n_2. \qquad ...(i)$$

Similarly, the pole of the line $l_2x + m_2y + n_2 = 0$ is

$$\left(-\frac{a^2l_1}{n_1}, -\frac{b^2m_1}{n_1}\right).$$

It will lie on the line $l_1x + m_1y + n_1 = 0$ if

$$\left(-\frac{a^2l_2}{n_2}\right)l_2+\left(-\frac{b^2m_2}{n_2}\right)$$

$$\Rightarrow \text{if} \quad a^2l_1l_2 + b^2m_1m_2 = n_1n_2$$

which is true because of (i) result. These two lines are called conjugate lines.

CHORD OF THE ELLIPSE WITH A GIVEN MIDDLE POINT

To find the equation of the chord of the ellipse

$\frac{x^2}{a^2} + \frac{y^2}{b^2} = 1$, *when its middle point is given.*

Let the equation of ellipse be $\frac{x^2}{a^2} + \frac{y^2}{b^2} = 1$,

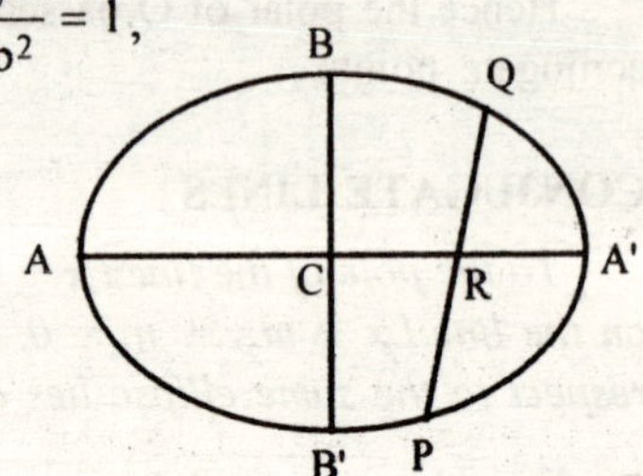

whose middle point R(x_1, + y_1) is given. Suppose (x'. y') and (x", y") be the co-ordinates of the point P and Q respectively.

Then we have

$$2x_1 = x' + x''$$

$$2y_1 = y' + y''$$

Equation of the chord PQ is given by

$$y - y_1 = \frac{y'' - y'}{x'' = x'} (x - x_1). \quad ...(i)$$

P, Q lie on the ellipse then we have

$$\frac{x'^2}{a^2} \frac{y'^2}{b^2} = 1 \text{ and } \frac{x''^2}{a^2} + \frac{y''^2}{b^2} = 1.$$

Subtracting, we have

$$\frac{x'^2 - x''^2}{a^2} + \frac{y'^2 - y''^2}{b^2} = 0$$

$$\Rightarrow \quad \frac{(y' + y'')(y' = y'')}{b^2} = -\frac{(x' + x'')(x' = x'')}{a^2}$$

$$\frac{y'' - y'}{x'' - x'} = -\frac{b^2}{a} - \frac{(x' + x'')}{(y' + y'')} = \frac{-b^2 x_1}{a^2 y_1}$$

Substituting $\frac{y'' - y'}{x'' - x'}$, (i) becomes

$$y - y_1 = -\frac{b^2 x_1}{a^2 y_1} (x - x_1)$$

$$\Rightarrow \quad yy_1 - y_1^2 = -\frac{b^2}{a^2} (xx_1 - x_1^2)$$

$$\Rightarrow \quad \frac{yy_1}{b^2} - \frac{y_1^2}{b^2} = -\frac{xx_1}{a^2} + \frac{x_1^2}{a^2}$$

$$\Rightarrow \quad \frac{xx_1}{a^2} - \frac{yy_1}{b^2} = \frac{x_1^2}{a^2} + \frac{y_1^2}{b^2}$$ which is the required equation.

TO FIND THE LOCUS OF THE MIDDLE POINT OF THE SYSTEM OF PARALLEL CHORDS

Let m be the slope of the system of parallel chord of the system. The equation of the chord is given as

$$\frac{xh}{a^2}+\frac{yk}{b^2}=\frac{h^2}{a^2}+\frac{k^2}{a^2}$$

The slope of this chord is $\dfrac{-b^2h}{a^2k}$.

But the slope is given to be m

$$\therefore \qquad m = + \frac{-b^2h}{a^2k}$$

$$\Rightarrow \qquad k = -\frac{-b^2}{a^2m}\ h.$$

Hence the locus of the point (h, k) is given by

$$y = -\frac{-b^2}{a^2m}\ x.$$

This is the diameter which bisect QR and chords which are parallel to at.

CONJUGATE DIAMETERS

Definition : *Two diameters are said to be conjugate which each bisects the chord parallel to other.*

Let $\qquad y = m_1x.$ ...(i)

and $\qquad y = m_2x.$ ...(ii)

be two conjugate diameters of the ellipse $\dfrac{x^2}{a^2}+\dfrac{y^2}{b^2}=1$. We know that the equation of the diameter bisecting the chords parallel to the diameter having slope m_1 is

$$y = \frac{-b^2}{a^2m_1}x. \qquad \text{...(iii)}$$

Since (iii) represents the diameter which bisects the chords parallel to the diameter, therefore the slopes of equation (ii) and (iii) must be identical. Hence we have

$\therefore \qquad m = \dfrac{-b^2}{a^2 m_1}$

or $\qquad m_1 m_2 = \dfrac{-b^2}{a^2}$

which is the required condition.

PROPERTIES OF CONJUGATE DIAMETERS

(a) The product of the focal distances of an extremity of a semidiameter of an ellipse is equal to the squares of the conjugate diameters.

Let CP and CD be two conjugate diameters of the ellipse $\dfrac{x^2}{a^2} + \dfrac{y^2}{b^2} = 1.$

If the co-ordinates of P is $(a \cos \theta, b \sin \theta)$ then the co-ordinates of D is $(-a \sin \theta, b \cos \theta)$. Let S and S' be two foci of the ellipse. Since P is the point on the ellipse,

$\therefore \qquad SP = a + ae \cos \theta, \; S'P = a - ae \cos \theta$

i.e., $\qquad SP.\ SP = (a + ae \cos \theta)(a - ae \cos \theta)$

$= a^2 - a^2e^2 \cos^2\theta$

$= a^2 - (a^2 - b^2) \cos^2 \theta$

$= a^2 \sin^2\theta + b^2 \cos^2 \theta = CD^2$

= square of the semi conjugate diameter.

(b) The area of the parallelogram formed by the tangents at the extremities of two conjugate diameter of an ellipse is constant and equal to the product of the axis.

Let PCP' and DCD' be two conjugate diameters. Tangents at P and P' are parallel to DCD' and tangents at D and D' are parallel to PCP'.

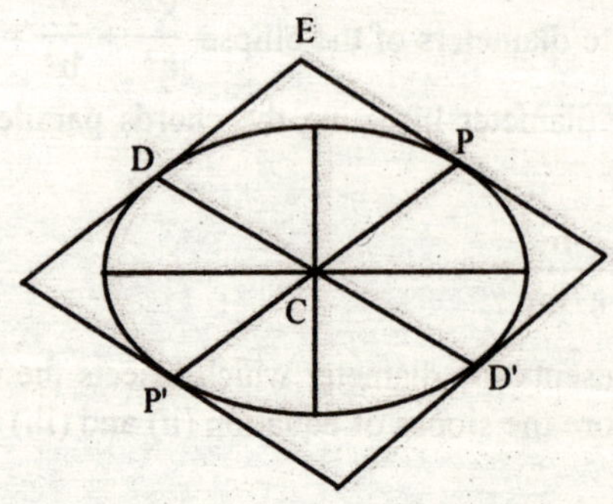

$\therefore$ Tangents at P, D; P', D' form a parall logram.

Area of the parallelogram

$$= 4 \text{ area of CPED}$$

$$= 8 \text{ area of } \Delta CPD$$

$$= 8 \times \frac{1}{2} \begin{vmatrix} 1 & 0 & 0 \\ 1 & a\cos\theta & b\cos\theta \\ 1 & -a\sin\theta & b\cos\theta \end{vmatrix}$$

$$= 4ab$$

$$= (2a) \times (2b)$$

= product of the axis.

(c) Show that tangents at the extremities of the diameter of an ellipse are parallel to its conjugate diameter.

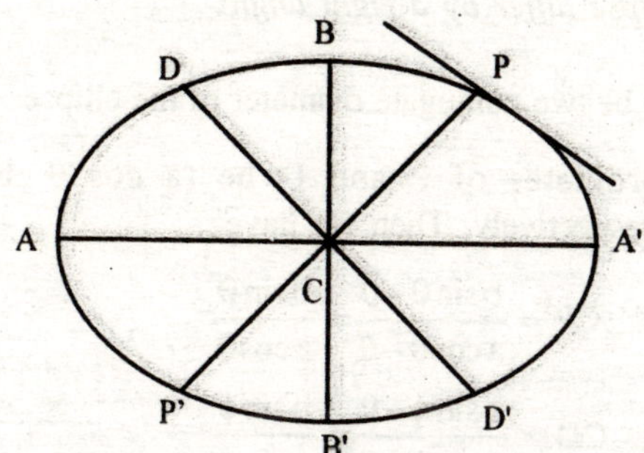

Let PCP' and DCD' be the conjugate diameters of the ellipse. Let the co-ordinates of P be (a cos θ, b sin θ), then the co-ordinates of D are (– a sin θ, b cos θ).

Equation of tangent at P is given as

$$\frac{x\cos\theta}{a} + \frac{y\sin\theta}{b} = 1$$

The slope of this tangent is $\frac{-b\cos\theta}{\sin\theta}$, which is same as slope of CD.

Therefore CD is parallel to the tangent at P. Similarly, it can be shown that CD is parallel to tangent at P' and CP is parallel to the tangents at D and D'

(d) Show that if CP and CD be two conjugate diameters of the ellipse $\frac{x^2}{a^2}+\frac{y^2}{b^2}=1$*, then* $CP^2 + CD^2 = a^2 + b^2$

Or

Show that the sum of the square of two conjugate semi-diameters is constant and equal to the sum of the squares of the semi-axis of the ellipse.

Let the co-ordinates of P be (a cos θ, b sin θ), then the co-ordinates of D is (–a sin θ, b cos θ). The co-ordinates of C is (0, 0)

$$\therefore \quad CP^2 = a^2 \cos^2\theta + b^2 \sin^2\theta$$

$$CD^2 = a^2 \sin^2\theta + b^2 \cos^2\theta.$$

Adding $CP^2 = CD^2 = a^2 + b^2$ which is constant and is equal to the sum of the squares of the semi-axis of the ellipse.

(e) Show that the eccentric angle of the extremities of a pair of conjugate diameters of an ellipse differ by a right angle.

Let CP and CD be two conjugate diameter of the ellipse $\frac{x^2}{a^2}+\frac{y^2}{b^2}=1$.

Suppose the co-ordinates of P and D be (a cos θ, b sin θ) and (a cos ϕ, b sin ϕ) respectively. Then we have

$$\therefore \text{ Slope of } \quad CP = \frac{b\sin\theta - 0}{a\cos\theta - 0} = \frac{b\sin\theta}{a\cos\theta}$$

$$\text{and slope of } \quad CD = \frac{b\sin\phi - 0}{a\cos\phi - 0} = \frac{b\sin\phi}{a\cos\phi}$$

Since the diameters are conjugate, therefore the product of their slopes is $-b^2/a^2$. Then we have

$$\therefore \quad \frac{b\sin\theta}{a\cos\theta}\cdot\frac{b\sin\phi}{a\cos\phi} = -\frac{b^2}{a^2}$$

$$\Rightarrow \quad \sin\theta\sin\phi + \cos\theta\cos\phi = 0$$

$$\Rightarrow \quad \cos(\theta - \phi) = 0 \Rightarrow \theta - \phi = \pm \pi/2.$$

Hence $\theta \sim \phi = \pi/2$.

Similarly, if PCP' and DCD' be two conjugate diameters then as summing the eccentric angle of P be θ the eccentric angle of D, P', D' are $\frac{\pi}{2}+\theta, \pi+\theta, \frac{3\pi}{2}+\theta$ respectively. In other words, the co-ordinates of D, P', D' are respectively (–a sin θ, b cos θ), (–a cos θ, –b sin θ), (–a sin θ, –b cos θ).

EQUI-CONJUGATE DIAMETERS

Definition: Two conjugate diameters of an ellipse, equal to each other are called Equi-conjugate diameters.

To find the equation of equi-cojugate diameters of an ellipse.

Let CP and CD be two conjugate diameters of the ellipse $\frac{x^2}{a^2} + \frac{y^2}{b^2} = 1$.

If the co-ordinates of P is $(a \cos \theta, b \sin \theta)$, then the co-ordinate of D is $(-a \sin \theta, b \cos \theta)$.

This give $CP^2 = a^2 \cos^2 \theta + b^2 \sin^2 \theta$

$$CD^2 = a^2 \sin^2 \theta + b^2 \cos^2 \theta.$$

For these conjugate diameters to be equi-conjugate diameters, then we have

$$CP^2 = CD^2$$

i.e. $\quad a^2 \cos^2 \theta + b^2 \sin^2 \theta = a^2 \sin^2 \theta + b^2 \cos^2 \theta$

$\Rightarrow \quad a^2 \cos 2\theta = b^2 \cos 2\theta.$

This gives $\cos 2\theta = 0 \Rightarrow \theta = \frac{\pi}{4}$ or $\frac{3}{4}\pi$.

If $\theta = \frac{\pi}{4}$, the equation of CP is given as

$$y = \frac{b \sin\theta}{a \cos\theta} x = \frac{b}{a} x$$

and the equation of CD is given as

$$y = \frac{-b \cos\theta}{a \sin\theta} x = \frac{-b}{a} x.$$

Hence the equations of the equi-conjugate diameters are

$$\frac{x}{a} = \pm \frac{y}{b}.$$

The other value of $\theta = \frac{3}{4} \pi$ will also give the same equation

A CIRCLE AND AN ELLIPSE

Show that a circle cuts an ellipse in four points and the sum of their eccentric angles is an even multiple of π radians.

Let the equ- of ellipse be

$$\frac{x^2}{a^2} + \frac{y^2}{b^2} = 1 \qquad \text{...(i)}$$

and the circle be

$$x^2 + y^2 + 2gx + 2fy + c = 0. \qquad \text{...(ii)}$$

Any point P on (i) is P (a cos θ, b sin θ).

This point lies on (ii)

$$\therefore \quad a^2 \cos^2 \theta + b^2 \sin^2 \theta + 2ga \cos \theta + 2fb \sin \theta + c = 0.$$

Using the relation,

$$\cos\theta = \frac{1-t^2}{1+t^2} \text{ and } \sin\theta = \frac{2t}{1+t^2}, \text{ where } t = \tan\frac{\theta}{2} \text{ we have}$$

$$a^2\left(\frac{1-t^2}{1+t^2}\right)^2 + b^2\left(\frac{2t}{1+t^2}\right)^2 + 2ga\left(\frac{1-t^2}{1+t^2}\right) + 2fb\left(\frac{2t}{1+t^2}\right) + c = 0$$

$$\Rightarrow \quad a^2(1-t^2)^2 + 4b^2t^2 + 2ga(1-t^2)(1+t^2) + 4fbt(1+t^2) + c(1-t^2)^2 = 0$$

$$\Rightarrow \quad (a^2 + c - 2ag)t^4 + 4bft^3 - 2(a^2 - 2b^2 - c)t^2 + 4bft + (a^2 + c + 2ga) \qquad \text{...(iii)}$$

This equation is biquadratic in t and therefore, it has four roots real or imaginary. Hence a circle cuts the ellipse in four points real or imaginary.

Let the roots of equation (iii) be given as

$$\tan\frac{\theta_1}{2}, \tan\frac{\theta_2}{2}, \tan\frac{\theta_3}{2}, \tan\frac{\theta_4}{2}, \tan$$

$$S_1 = \Sigma\tan\frac{\theta_1}{2} = \frac{-4bf}{a^2 + c - 2ga}$$

$$S_3 = \Sigma\tan\frac{\theta_1}{2}\tan\frac{\theta_2}{2}\tan\frac{\theta_1}{2} = \frac{-4bf}{a^2 + c^2 - 2ga}$$

$$\therefore \quad \tan\left(\frac{\theta_1}{2} + \frac{\theta_2}{2} + \frac{\theta_3}{2} + \frac{\theta_4}{2}\right) = \frac{S_1 - S_3}{1 - S_2 + S_4} = 0.$$

$$\Rightarrow \quad \frac{\theta_1}{2} + \frac{\theta_2}{2} + \frac{\theta_3}{2} + \frac{\theta_4}{2} = n\pi$$

or $\qquad \theta_1 + \theta_2 + \theta_3 + \theta_4 = n\pi,$

which is an even multiple of π.

Example:

Show that the locus of the middle points of the chords of the ellipse

$$\frac{x^2}{a^2} + \frac{y^2}{b^2} = 1$$

which touch a concentric circle of radius c is the curve

$$c^2\left(\frac{x^2}{a^4} + \frac{y^2}{b^4}\right) = \left(\frac{x^2}{a^2} + \frac{y^2}{b^2}\right)^2$$

Solution:

The equation of the ellipse is

$$\frac{x^2}{a^2} + \frac{y^2}{b^2} = 1. \qquad \text{...(i)}$$

The equation of the chord with middle point (h, k) is

$$\frac{xh}{a^2} + \frac{yk}{b^2} = \frac{h^2}{a^2} + \frac{k^2}{b^2}. \qquad \text{...(ii)}$$

Since the centre of the ellipse is (0, 0), therefore, the centre of the concentric circle is also (0, 0). Now (ii) touches the concentric circle if the perpendicular distance from the centre of the circle to the line (ii) is equal to the radius c.

i.e.
$$\frac{\left(\frac{h^?}{a^2} + \frac{k^2}{b^2}\right)}{\sqrt{\frac{h^2}{a^4} + \frac{k^2}{b^4}}} = c$$

$$\Rightarrow \qquad \left(\frac{h^2}{a^2} + \frac{k^2}{b^2}\right) = c^2\left(\frac{h^2}{a^4} + \frac{k^2}{b^4}\right).$$

Hence the locus of (h, k) is given as

$$\left(\frac{x^2}{a^2} + \frac{y^2}{b^2}\right) = c^2\left(\frac{x^2}{a^4} + \frac{y^2}{b^4}\right).$$

MISCELLANEOUS EXAMPLES

Example 1:

Find the equation to the ellipses, whose centres are the origin, whose axes are the axes of co-ordinates, and which pass through (α) the points (2, 2) and (3, 1) and (β) the points (1, 4) and (–6, 1)

Solution:

Let the equation to any ellipse be

$$\frac{x^2}{a^2} + \frac{y^2}{b^2} = 1 \qquad ...(1)$$

(a) As the points (2, 2) and (3, 1) lie on (1), we have

$$\frac{4}{a^2} + \frac{4}{b^2} = 1 \qquad ...(2)$$

and $$\frac{9}{a^2} + \frac{1}{b^2} = 1 \qquad ...(3)$$

Solving (2) and (3), we get $a^2 = \frac{32}{3}$ and $b^2 \frac{1}{5}$

Putting in (1), the equation to the ellipse will be

$$\frac{3x^2}{32} + \frac{5y^2}{32} = 1 \quad \text{or} \quad 3x^2 + 5y^2 = 32.$$ **Ans.**

(b) As the points (1, 4) and (–6, 1) lie on (1), we have

$$\frac{1}{a^2} + \frac{16}{b^2} = 1 \qquad ...(2)$$

and $$\frac{36}{a^2} + \frac{1}{b^2} = 1 \qquad ...(3)$$

Solving (2) and (3), we get $a^2 = \frac{115}{3}$ and $b^2 = \frac{115}{7}$

Putting in (1) we have

$$\frac{3x^2}{115} + \frac{7y^2}{115} = 3x^2 + 7y^2 = 115$$ **Ans.**

Example 2:

If ϕ, ϕ' be the angles subtended by the major axis of an ellipse at the extremities of a pair of conjugate diameters, show that $\cot^2\phi + \cot^2\phi'$ is constant.

Solution:

Let the extremities of two conjugate semi diameters be p(a cos θ, b sinθ) and D (–a sin θ, b cos θ). Also the co-ordinates of A and A' are (a, 0) and (–a, 0).

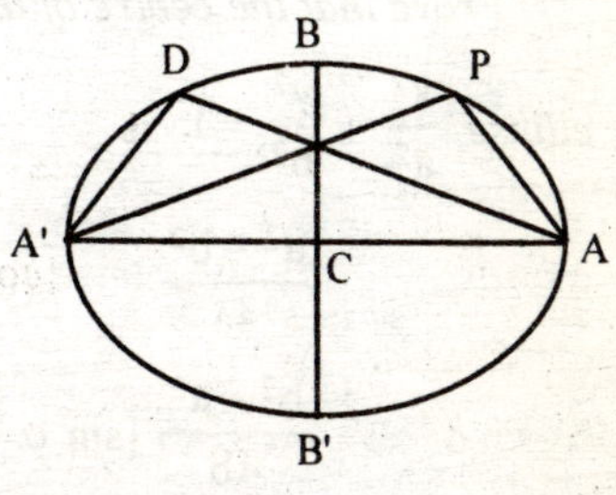

$$\text{slope of } AP = \frac{b\sin\theta - 0}{a\cos\theta - a}$$

$$= \frac{b.2\sin\theta/2\cos\theta/2}{-a2\sin^2\theta/2} = -\frac{b}{a}\cot\frac{\theta}{2}$$

$$\text{slope of } A'P = \frac{b\sin\theta}{a\cos\theta + a} = \frac{b.2\sin\theta/2\cos\theta/2}{2a\cos^2\theta/2} = \frac{b}{a}\tan\frac{\theta}{2}.$$

Since ϕ is the angle between AP and A'P.

$$\therefore \quad \tan\phi = \frac{\frac{b}{a}\tan\frac{\theta}{2} + \frac{b}{a}\cot\frac{\theta}{2}}{1 - \frac{b}{a}\tan\frac{\theta}{2}\cot\frac{\theta}{2}}$$

$$= \frac{b}{a}\left[\frac{\sin\frac{\theta}{2}}{\cos\frac{\theta}{2}} + \frac{\cos\frac{\theta}{2}}{\sin\frac{\theta}{2}}\right]\left[\frac{1}{1 - \frac{b^2}{a^2}}\right]$$

$$= \frac{ab\left(\sin\frac{\theta}{2} + \cos^2\frac{\theta}{2}\right)}{\left(a^2 - b^2\right)\sin\frac{\theta}{2}\cos\frac{\theta}{2}}$$

Similarly, $\tan\phi' = \dfrac{2ab}{\left(a^2 - b^2\right)\cos\theta}$

$$\therefore \cot^2\phi + \cot^2\phi' = \frac{\left(a^2 - b^2\right)^2}{4a^2b^2}\left(\sin^2\theta + \cos^2\theta\right)$$

$$= \frac{\left(a^2 - b^2\right)^2}{4a^2b^2}$$

$$= \text{constant.}$$

Example 3:

Prove that the centre of the circle passing through points α, β, γ on the ellipse $\frac{x^2}{a^2}+\frac{y^2}{b^2}=1.$ is

$$\frac{a^2-b^2}{4a}\{\cos\alpha+\cos\beta+\cos\gamma+\cos(\alpha+\beta+\gamma)\},$$

$$\frac{b^2-a^2}{4b}\{\sin\alpha+\sin\beta+\sin\gamma+\sin(\alpha+\beta+\gamma)\}.$$

Solution:

We have that circle and ellipse intersect at four points.

Let the fourth point be d. Then we have

$$\alpha+\beta+\gamma+\delta=2n\pi. \qquad ...(i)$$

Let the equation of the circle be

$$x^2+y^2+2gx+2fy+c=0. \qquad ...(ii)$$

Substituting $x = a\cos\theta$, $y = b\sin\theta$, we get

$$a^2\cos^2\theta+b^2\sin^2\theta+2ga\cos\theta+2fb\sin\theta+c=0. \qquad ...(iii)$$

$$\Rightarrow a^2\cos^2\theta+b^2(1-\cos^2\theta)+2ga\cos\theta+c-2fb\sqrt{1-\cos^2\theta}$$

$$\Rightarrow \{a^2\cos^2\theta+b^2(1-\cos^2\theta)+2ga\cos\theta+c\}^2-4f^2b^2(1-\cos^2\theta).$$

This equation is fourth degree equation in $\cos\theta$, hence it roots are $\cos\alpha$, $\cos\beta$, $\cos\gamma$, $\cos\delta$.

$$\therefore \quad \cos\alpha\cos\beta\cos\gamma\cos\delta=-\frac{\text{coeff. of }\cos^3\theta}{\text{coeff. of }\cos^4\theta}$$

$$=\frac{4ga(a^2-b^2)}{(a^2-b^2)^2}=-\frac{4ga}{a^2-b^2}. \qquad ...(iv)$$

Similarly, changing (iii) in terms of $\sin\theta$, we have

$$\sin\alpha+\sin\beta+\sin\gamma+\sin\delta=\frac{4fb}{b^2-a^2} \qquad ...(v)$$

The centre of circle (ii) is the point $(-g, -f)$, whose values are given by (iv) and (v).

∴ The centre is given as

$$-g = \frac{a^2 - b^2}{4a} (\cos\alpha + \cos\beta + \cos\gamma + \cos\delta), \qquad \text{...(vi)}$$

$$-f = \frac{b^2 - a^2}{4a} (\sin\alpha + \sin\beta + \sin\gamma + \sin\delta), \qquad \text{...(vii)}$$

But we have $\quad \delta = 2n\pi - \sqrt{\alpha + \beta + \gamma}$

$\therefore\ \cos\alpha = \cos(\alpha + \beta + \gamma)$, sind $= -\sin(\alpha + \beta + \gamma)$

Substituting the values, (vi) and (vii) gives the required result.

Example 4:

Show that the locus of the middle point of the chords of the ellipse $\frac{x^2}{a^2} + \frac{y^2}{b^2} = 1$ *which subtend a right angle at the centre is*

$$\frac{x^2}{a^4} + \frac{y^2}{b^4} = \frac{a^2 + b^2}{a^2 b^2}\left(\frac{x^2}{a^2} + \frac{y^2}{b^2}\right)^2.$$

Solution:

The equation of the chord with (x_1, y_1) middle point is given as

$$\frac{xx_1}{a^2} + \frac{yy_1}{b^2} = \frac{x_1^2}{a^2} + \frac{y_1^2}{a^2}. \qquad \text{...(i)}$$

Making the equation of the given ellipse homogenous with the help of (i), we have

$$\frac{x^2}{a^2} + \frac{y^2}{b^2} = \left\{\left(\frac{xx_1}{a^2} + \frac{yy_1}{b^2} + \right) \Big/ \left(\frac{x_1^2}{a^2} + \frac{y_1^2}{a^2}\right)\right\}^2$$

$$\left(\frac{x^2}{a^2} + \frac{y^2}{b^2}\right)\left(\frac{x_1^2}{a^2} + \frac{y_1^2}{b^2}\right)^2 - \left(\frac{xx_1}{a^2} + \frac{yy_1}{b^2}\right)^2 = 0.$$

This is the joint equation of the straight line joining the centre to the extremities of the chord. Since these straight lines are perpendicular to each other, we must have coefficient of x^2 + coefficient of $y^2 = 0$

i.e. $$\left\{\frac{1}{a^2}\left(\frac{x_1^2}{a^2} + \frac{y_1^2}{b^2}\right)^2 - \frac{x_1^2}{a^4}\right\} + \left\{\frac{1}{b^2}\left(\frac{x_1^2}{a^2} + \frac{y_1^2}{b^2}\right)^2 - \frac{x_1^2}{b^4}\right\} = 0$$

$$\Rightarrow \quad \frac{x_1^2}{a^4} + \frac{y_1^2}{b^4} = \left(\frac{1}{a^2} + \frac{1}{b^2}\right)\left(\frac{x_1^2}{a^2} + \frac{y_1^2}{b^2}\right)^2.$$

∴ Locus of (x_1, y_1) is

$$\frac{x^2}{a^4} + \frac{y^2}{b^4} = \left(\frac{a^2 + b^2}{a^2 b^2}\right)\left(\frac{x^2}{a^2} + \frac{y^2}{b^2}\right)^2.$$

Example 5:

With a given point and line as focus and directrix, a series of ellipse are described; prove that the locus of the extremities of their minor axes is a parabola.

Solution:

Let S be the given focus and ZM be the given line. Then we have

$SZ = CZ - CS = a/e - ae$

$= (a/e)(1 - e^2) = b^2/ae = k$ (say) (constant)

as $\quad b^2 = a^2(1 - e^2)$

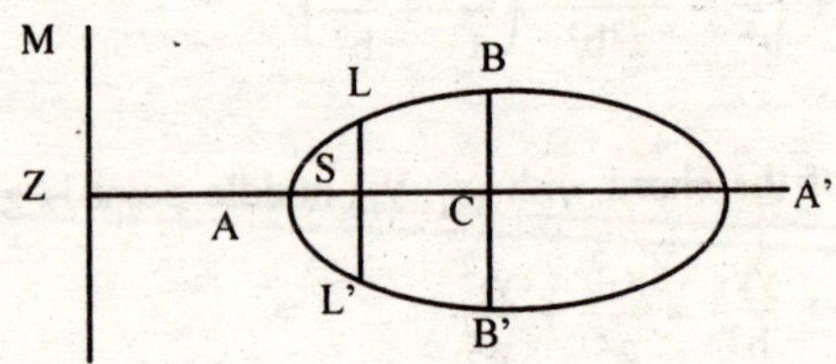

Now taking SC as axis of x and LSL', as y-axis.

Let (x, y) be the co-ordinates of B with respect to these axes; then

$x = SC = ae,$

and $\quad y = CB = b;$

hence $\quad y^2/x = b^2/ae = k.$

Therefore $y_2 = kx$, is the required locus and is clearly a parabola.

Example 6:

A line of fixed length a + b moves so that its ends are always on two fixed perpendicular straight lines; prove that the locus of a point, which divides this line into portions of length a and b, is an ellipse.

Solution:

Suppose APB is the line which moves with its ends B and A on x-axis and y-axis respectively and let AP = a, PB = b, so that AB = a + b.

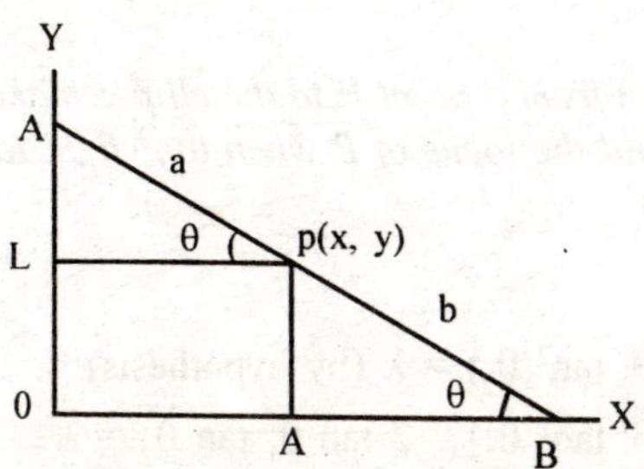

If AB make an angle θ with the axis of x and the co-ordinates of P be (x, y); then

from ΔAPL, $x = a\cos\theta$ ...(1)

and from ΔPBQ, $y = b\sin\theta$ ...(2)

Eliminating Q from (1) and (2),

we get $x^2/a^2 + y^2/b^2 = 1$ which is an ellipse **Proved.**

Example 7:

From any point P on the ellipse, PN is drawn perpendicular to the axis and produced to Q, so that NQ equals PS, where S is a focus; prove that locus of Q is the two straight lines $y \pm ex + a = 0$.

Solution:

Let the co-ordinates of Q be (x, y), then we have

$x = CN = x1$ ...(1)

and $y = NQ = (a + ex1)$ ...(2)

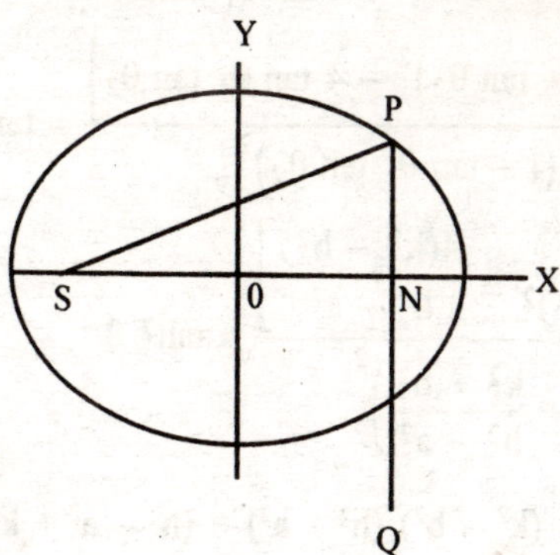

(2), the required locus is $y = -(a + ex)$ or $y + a + ex = 0$

Taking S' instead of S, the locus of Q will be $y + a - ex = 0$.

Proved.

Example 8:

The tangents drawn from a point P to the ellipse make angle θ_1 and θ_2 with the major axis; find the locus of P when $\tan^2 \theta_1 + \tan^2 \theta_2$ is constant $(= \lambda)$.

Solution:

$$(\tan^2 \theta_1 + \tan^2 \theta_2) = \lambda \text{ (by hypothesis)}$$

hence $(\tan^2 \theta_1 + \tan^2 \theta_2) - 2 \tan \theta_1 \tan \theta_2 = \lambda$

$$\Rightarrow \left(\frac{2hk}{h^2 - a^2}\right)^2 - \frac{2(k^2 - b^2)}{h^2 - a^2} = \lambda$$

$$\Rightarrow 4h^2k^2 - 2(k^2 - b^2)(h^2 - a^2) = \lambda (h^2 - a^2)^2$$

$$\Rightarrow 2(h^2k^2 + k^2a^2 + h^2b^2 - a^2b^2 = \lambda (h^2 - a^2)^2$$

Generalising the locus of (h, k) is

$$2(x^2y^2 + a^2y^2 + x^2b^2) = \lambda (x^2 - a^2)^2.$$ **Ans.**

Example 9:

Find the locus of the intersection of tangents which meet at a given angle α.

Solution:

Let θ_1 and θ_2 be the angles of inclination of tangents through P, with the x-axis, then $\theta_1 - \theta_2 = \alpha$ by hypothesis.

So $\tan(\theta_1 - \theta_2) = \tan a$ or $\dfrac{\tan \theta_1 - \tan \theta_2}{1 + \tan \theta_1 \tan \theta_2} = \tan \alpha$

$$\Rightarrow \frac{\left[(\tan \theta_1 + \tan \theta_2)^2 - 4 \tan \theta_1 \tan \theta_2\right]}{(1 + \tan \theta_1 \tan \theta_2)^2} = \tan^2 \alpha$$

$$\Rightarrow \frac{\left[\frac{4h^2k^2}{(h^2 - a^2)^2} - \frac{4(k^2 - b^2)}{h^2 - a^2}\right]}{\left(1 + \frac{k^2 - b^2}{h^2 - a^2}\right)^2} = \tan^2 \theta$$

$$\Rightarrow 4h^2k^2 - 4(k^2 - b^2)(h^2 - a^2) = (h^2 - a^2 + k^2 - b^2) \tan^2 \alpha$$

$$\Rightarrow 4(b^2h^2 + a^2k^2 - a^2b^2) = (h^2 + k^2 - a^2 - b^2) \tan^2 \alpha.$$

Generalising, the locus of the point of intersection of tangents is $4(b^2x^2 + a^2y^2 - a^2b^2) = (x^2 + y^2 - a^2 - b^2) \tan^2 \alpha$.

Example 10(a):

Find the locus of the intersection of tangents if the lines joining the points of contact to the centre be perpendicular.

Solution:

If (h, k) be the points of intersection of two tangents of the ellipse

$$x^2/a^2 + y^2/b^2 = 1 \quad \text{...(1)}$$

Then equation to the chord of contact of (h, k) w.r.t. (1) is given by

$$\frac{hx}{a^2} + \frac{ky}{b^2} = 1 \quad \text{...(2)}$$

The equation to the lines joining the points of contact to the centre will be obtained by making (1) homogeneous with (2). So, we get

$$\frac{x^2}{a^2} + \frac{y^2}{b^2} = \left(\frac{hx}{a^2} + \frac{ky}{b^2}\right)^2$$

$$x^2\left(\frac{h^2}{a^4} - \frac{1}{a^2}\right) + y^2\left(\frac{k^2}{b^4} - \frac{1}{b^2}\right) + \frac{2xyhk}{a^2b^2} = 0 \quad \text{...(3)}$$

The lines given by (3) will be perpendicular to each other if coefficient of x^2 + coefficient of $y^2 = 0$.

i.e.
$$\frac{h^2}{a^4} - \frac{1}{a^2} + \frac{k^2}{b^4} - \frac{1}{b^2} = 0.$$

Generalising, the locus of point (b, k) is given by

$$\frac{x^2}{a^4} + \frac{y^2}{b^4} = \frac{1}{a^2} + \frac{1}{b^2}$$

Ans.

Example 10(b):

Find the locus of the middle points of chords of an ellipse which subtend a right at the centre.

Solution:

Let the equation the ellipse be,

$$x^2/a^2 + y^2/b^2 = 1 \quad \text{...(1)}$$

and the co-ordinates of the mid-point of the chord which subtends a right angle at the centre be (h, k).

The equation of the chord will be

$$\frac{xh}{a^2} + \frac{yk}{b^2} = \frac{h^2}{a^2} + \frac{k^2}{b^2}$$

$$\Rightarrow \quad \left(\frac{xh}{a^2}+\frac{yk}{b^2}\right)\Big/\left(\frac{h^2}{a^2}+\frac{k^2}{b^2}\right)=1. \qquad \ldots(2)$$

The equation to the lines joining the origin to the points of intersection of (1) and (2) will be obtained by making (1) homogeneous with the help of (2). Hence the equation is

$$\frac{x^2}{a^2}+\frac{y^2}{b^2}=\left\{\left(\frac{xh}{a^2}+\frac{yk}{b^2}\right)\left(\frac{h^2}{a^2}+\frac{k^2}{b^2}\right)\right\}^2$$

$$\Rightarrow \quad (h^2x^2+a^2y^2)(h^2b^2+a^2k^2)=a^2b^2(b^2xh+a^2yk)^2. \qquad \ldots(3)$$

If the lines represented by (3) are at right angles, the sum of the coefficients of x^2 and y^2 must by zero. Simplifying (3) and equating the sum of the coefficient in zero, we get

$$(a^2+b^2)(b^2h^2+a^2k^2)^2=a^2b^2(b^2h^2+a^2k^2)^2=0$$

Generalising we get the required locus as

$$(a^2+b^2)(b^2x^2+a^2y^2)^2=a^2b^2(b^4x^2+a^4y^2)^2. \qquad \textbf{Ans.}$$

Example 10(c):

Find the locus of the middle points of chords of an ellipse whose poles are on the auxiliary circle.

Solution:

The equation to the chord having (h, k) as middle point is $\frac{xh}{a^2}+\frac{yk}{b^2}$ $\frac{h^2}{a^2}+\frac{k^2}{b^2}$. If (α, β) be the pole of this chord with respect to the ellipse $x^2/a^2+y^2/b^2=1$, then the polar of the point is $x\alpha/a^2+y\beta/b^2=1$.

By hypothesis these two lines are same; so on comparing the coefficients, we have $\frac{\alpha}{h}=\frac{\beta}{k}=\frac{1}{h^2/a^2+k^2/b^2}$. whence

$$\alpha=\frac{h}{h^2/a^2+k^2/b^2}=\frac{a^2b^2h}{b^2h^2+k^2k^2} \text{ and } \beta=\frac{a^2b^2k}{b^2h^2+k^2k^2}$$

As the point (α, β) lies on the auxiliary circle $x^2+y^2=a^2$, we must have

$$\frac{a^4b^4h^2}{(b^2h^2+a^2k^2)}+\frac{a^4b^4k^2}{(b^2h^2+a^2k^2)}$$

or $\quad a^2b^4(h^2+k^2)=(b^2h^2+a^2k^2)^2$

Generalising, the locus of (h, k) is $a^2b^4 (x^2 + y^2) = (b^2x^2 + a^2y^2)^2$. **Ans.**

Example 11:

Show that the locus of the pole, with respect to the auxiliary circle, of a tangent to the ellipse is a similar concentric ellipse, whose major axis is at right angles to that of the original ellipse.

Solution:

The equation to tangent at any point (a cos f, b sin f) of the ellipse

$$\frac{x^2}{a^2} + \frac{y^2}{b^2} = 1 \text{ will be } \frac{x}{a}\cos\phi + \frac{y}{b}\sin\phi = 1. \qquad ...(1)$$

Let (h, k) be the co-ordinates of the pole of this tangent w.r.t. the auxiliary circle $x^2 + y^2 = a^2$, then equation to its polar will be

$$xh + yk = a^2. \qquad ...(2)$$

As the equations (1) and (2) are identical so comparing

$$\frac{\cos\phi}{ah} = \frac{\sin\phi}{bk} = \frac{1}{a^2}; \text{ hence } \cos\phi = \frac{h}{a} \text{ and } \sin\phi = \frac{bk}{a^2}.$$

To eliminating ϕ, squaring and adding, we have

$$\cos^2\phi + \sin^2\phi = \frac{h^2}{a^2} + \frac{b^2k^2}{a^4} = 1.$$

Generalising the locus of (h, k) is $a^2x^2 + b^2y^2 = a^4$. **Ans.**

Clearly this represents an ellipse whose major axis is the y-axis. The axis of the original ellipse was x-axis, hence the two are perpendicular to each other.

Example 12:

The tangents drawn from a point P to the ellipse make angles q_1 and q_2 with the major axis; find the locus of P when $q_1 + q_2$ is constant (=2a).

Solution:

Let the equation of the tangent to the ellipse $x^2/a^2 + x^2/b^2 = 1$ be $y = mx + \sqrt{(a^2m^2 + b^2)}$ and let the co-ordinates of the point P be (h, k). If (h, k) lies on the tangent, then $k = mh + \sqrt{(a^2m^2 + b^2)}$

or $(k - mh)^2 = a^2m^2 + b^2$ or $m^2(h^2 - a^2) - 2mhk + (k^2 - b^2) = 0$.

As this equation is quadratic in 'm', it will give two roots of m, say m_1 and m_2. These correspond to the slopes of tangents through the point P whose locus is required. Also, if θ_1 and θ_2 be angles of inclination of tangents with axis of x, then $\tan\theta_1 = m_1$ and $\tan\theta_2 = m_2$.

Hence $m_1 + m_2 = \dfrac{2hk}{h^2 - a^2}$ or $\tan\theta_1 + \tan\theta_2 = \dfrac{2hk}{h^2 - a^2}$ ...(1)

By hypothesis, $\theta_1 + \theta_2 = 2\alpha$;

so $\tan(\theta_1 + \theta_2) = \tan 2\alpha.$

$$\Rightarrow \quad \frac{\tan\theta_1 + \tan\theta_2}{1 - \tan\theta_1 \tan\theta_2} = \tan 2\alpha \text{ or } \frac{\dfrac{2hk}{h^2 - a^2}}{1 - \dfrac{k^2 - b^2}{h^2 - a^2}} = \tan 2\alpha.$$

By (1) and (2)

$$\Rightarrow \quad \frac{2hk}{h^2 - a^2 - k^2 + b^2} = \tan 2\alpha$$

$$\Rightarrow \quad h^2 - k^2 - a^2 + b^2 = 2hk \cot 2\alpha.$$

Hence the locus of $P = (h, k)$ is $x^2 - 2xy \cot 2\alpha - y^2 = a^2 - b^2$. **Ans.**

Example 13:

The eccentric angles of two points p and Q on the ellipse are ϕ_1 and ϕ_2; prove that the area of the parallelogram formed by the tangents at the ends of the diameters through P and Q is 4ab cosec $(\phi_1 - \phi_2)$ and hence that it is least when P and Q are at the end of conjugate diameters.

Solution:

Equations to the tangent at the points P and Q are

$$\frac{x}{a}\cos\phi_1 + \frac{y}{b}\sin\phi_2 = 1 \quad ...(1)$$

and

$$\frac{x}{a}\cos\phi_2 + \frac{y}{b}\sin\phi_2 = 1 \quad ...(2)$$

Solving (1) and (2), we will have the co-ordinates of L. Multiplying (1) by $\sin\phi_2$ and (2) by $\sin\phi_1$ and subtracting, we get

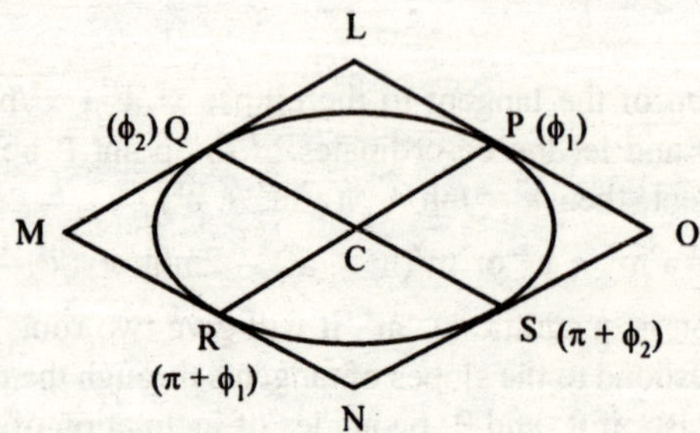

$$\frac{x}{a}(\sin\phi_2\cos\phi_1-\cos\phi_2\sin\phi_1)=\sin\phi_2-\sin\phi_1$$

$$\Rightarrow \quad \frac{x}{a}\sin(\phi_2-\phi_1)=2\sin\frac{\phi_2-\phi_1}{2}\cos\frac{\phi_2+\phi_1}{2}.$$

$$\therefore x=a\frac{\cos\frac{\phi_1+\phi_2}{2}}{\cos\frac{\phi_1-\phi_2}{2}};\ \text{similiarly } y=b\frac{\sin\frac{\phi_1+\phi_2}{2}}{\cos\frac{\phi_1-\phi_2}{2}}$$

Above are the co-ordinates of the point of intersection L.

Putting $\phi_1=\pi$ in above, we get the co-ordinates of the point of intersection M of tangent at Q and R as

$$\left\{\frac{a\sin\frac{\phi_1+\phi_2}{2}}{\sin\frac{\phi_1-\phi_2}{2}},\frac{-b\cos\frac{\phi_1+\phi_2}{2}}{\sin\frac{\phi_1-\phi_2}{2}}\right\}$$

Area of the parallelogram is 4 ΔCLM = $4\,\frac{1}{2}\,(x_1y_2 - x_2y_1)$

$$=2.\frac{ab}{\sin\frac{\phi_1-\phi_2}{2}\cos\frac{\phi_1-\phi_2}{2}}\left[-\cos\frac{2\,\phi_1+\phi_2}{2}-\sin\frac{2\,\phi_1-\phi_2}{2}\right]$$

$$=2.\frac{-4ab}{\sin\phi_1-\phi_2}\ 4b\operatorname{cosec}(\phi_1-\phi_2)$$

The area is least when $\sin(\phi_1-\phi_2)$ is greatest, i.e., when $\phi_1-\phi_2=\pi/2$ or when P and Q are the ends of conjugate diameters.

Example 14:

Circles of constant radius c are drawn to pass through the ends of a variable diameter of the ellipse. Prove that the locus of their centres is the curve $(x^2+y^2)(a^2x^2+b^2y^2+a^2b^2)=c^2(a^2x^2+b^2y^2)$.

Solution:

The co-ordinates of the ends of any diameters of the ellipse $x^2/a^2+y^2/b^2=1$ may be taken as

$$(a\cos\phi,\ b\sin\phi)\ \text{and}\ (-a\cos\phi,\ -b\sin\phi).$$

Again taking, (h, k) as the co-ordinates of the centre of the circle of given radius c; then equation tot he circle will be $(x - h)^2 + (y - k)^2 = c^2$. Since the circle passes through the ends of the given diameter, their co-ordinates must satisfy it.

So $$(a \cos \phi - h)^2 + (b \sin \phi - k)^2 = c^2 \quad \text{...(1)}$$

and $$(-a \cos \phi - h)^2 + (-b \sin \phi - k)^2 = c^2 \quad \text{...(2)}$$

Adding (1) and (2),

$$2(a^2 \cos^2 \phi + h^2) + 2(b^2 \sin^2 \phi - k^2) = 2c^2 \quad \text{...(3)}$$

and subtracting (1) from (2), $4ah \cos \phi + 4bk \sin \phi = 0$

$$\Rightarrow \quad \frac{-\cos \phi}{bk} = \frac{\sin \phi}{ah} = \frac{1}{\sqrt{(b^2k^2 + a^2h^2)}} \quad \text{...(4)}$$

Putting the values of $\sin \phi$ and $\cos \phi$ from (4) in (3), we get

$$\frac{a^2b^2(h^2 + k^2)}{b^2k^2 + a^2h^2} + h^2 + k^2 = c^2$$

$$\Rightarrow \quad (h^2 + k^2)(a^2b^2 + a^2h^2 + b^2k^2) + c^2 (a^2h^2 + b^2k^2) = 0$$

Generalising, the locus of the centre (h, k) is

$$(x^2 + y^2)(a^2x^2 + b^2y^2 + a^2b^2) = c^2 (a^2x^2 + b^2y^2).$$

Example 15:

The tangents drawn from a point P to the ellipse make angles θ_1 and θ_2 with the major axis; find the locus of P when $\tan \theta_1 + \tan \theta_2$ is constant (= c).

Solution:

By hypothesis, $\tan \theta_1 + \tan \theta_2 = c$ and

$$\tan \theta_1 + \tan \theta_2 = \frac{2hk}{h^2 - a^2}$$

So $\dfrac{2hk}{h^2 - a^2} = c$ or $2hk = c(h^2 - a^2)$

Generalising the locus of P (h, k) is $2xy = c(x^2 - a^2)$. **Ans.**

Example 16:

The tangents drawn from a point P to the ellipse make angles θ_1 and θ_2 with the major axis; find the locus of P when $\tan \theta_1 - \tan \theta_2$ is constant (= d).

Solution:

$\tan\theta_1 - \tan\theta_2 = d$ (by hypothesis) hence

$(\tan\theta_1 + \tan\theta_2)^2 - 4\tan\theta_1 \tan\theta_2 = (\tan\theta_1 - \tan\theta_2)2 = d^2$

$$\Rightarrow \quad \left(\frac{2hk}{h^2 - a^2}\right)^2 - \frac{4(k^2 - b^2)}{b^2 - a^2} = d^2$$

$$\Rightarrow \quad 4h^2k^2 - 4(h^2 - a^2)(k^2 - b^2) = d^2(h^2 - a^2)^2$$

$$\Rightarrow \quad 4(b^2h^2 + a^2k^2 - a^2b^2) = d^2(h^2 - a^2)^2.$$

Generalising, the locus of (h, k) is

$$4(b^2x^2 + a^2y^2 - a^2b^2) = d^2(x^2 - a^2)^2.$$ **Ans.**

$$\frac{x}{a}\sin\frac{\phi_1 + \phi_2}{2} - \frac{y}{b}\sin\frac{\phi_1 + \phi_2}{2} = 0 \quad \ldots(5)$$

Now, we have to eliminate ϕ_1 and ϕ_2 from (4) and (5).

So squaring (4) and (5) and adding, we have

$$\frac{x^2}{a^2}\left(\cos\frac{2\phi_1 + \phi_2}{2} + \sin\frac{2\phi_1 + \phi_2}{2}\right)$$

$$+ \frac{y^2}{b^2}\left(\sin\frac{2\phi_1 + \phi_2}{2} + \cos\frac{2\phi_1 + \phi_2}{2}\right) = 4$$

$\Rightarrow \quad x^2/a^2 + y^2/b^2 = 4$ which is the required locus. **Ans.**

Example 17:

Find the locus of the intersection of tangents of the sum of the eccentric angles of their points of contact be equal to a constant angle 2α.

Solution:

If ϕ_1 and ϕ_2 be the eccentric angles of the points of contact of tangents on the ellipse $(x^2/a^2) + (y^2/b^2) = 1$. ...(1)

Then $\phi_1 + \phi_2 = 2a$ (by hypothesis) or $\dfrac{\phi_1 + \phi_2}{2} = \alpha$ ...(2)

The tangent at point ϕ_1 i.e., $(a\cos\phi_1, b\sin\phi_1)$ to (1) is

$$(x/a)\cos\phi_1 \; (y/b)\sin\phi_1 = 1 \quad \ldots(3)$$

and tangent at point ϕ_2, i.e., $(a\cos\phi_2, b\sin\phi_2)$ to (1) is

$$(x/a)\cos\phi_2 + (y/b)\sin\phi_2 = 1 \quad \ldots(4)$$

Now, we have to eliminate ϕ_1 and ϕ_2 from (2), (3) and (4).

Subtracting (4) from (3), we have

$$\frac{x}{a}(\cos\phi_1 - \cos\phi_2) + \frac{y}{b}(\sin\phi_1 - \sin\phi_2) = 0$$

$$\Rightarrow \frac{x}{a}\left(2\sin\frac{\phi_1+\phi_2}{2}\sin\frac{\phi_1+\phi_2}{2}\right)+\frac{y}{b}\left(2\sin\frac{\phi_1+\phi_2}{2}\cos\frac{\phi_1+\phi_2}{2}\right)=0$$

$$\Rightarrow \frac{x}{a}\sin\frac{\phi_1+\phi_2}{2} - \frac{y}{b}\cos\frac{\phi_1+\phi_2}{2} = 2$$

Putting the value of $(\phi_1 + \phi_2)/2$ from (2), we get the locus of point of intersection of tangents as

$(x/a) \sin a - (y/b) \cos a = 0$ or $ay = bx \tan \alpha$. **Ans.**

Example 18:

Find the locus of the middle points of chorus of the ellipse which pass through the given point (h, k).

Solution:

If the middle point of the chord of the ellipse

$\frac{x^2}{a^2} + \frac{y^2}{b^2} = 1$ be (x_1, y_1),

the equation to this chord be

$$\frac{xx_1}{a^2} + \frac{yy_1}{b^2} = \frac{x_1^2}{a^2} + \frac{y_1^2}{b^2} \quad ...(1)$$

As (1) passes through (h, k), we have $\frac{hx_1}{a^2} + \frac{ky_1}{b^2} = \frac{x_1^2}{a^2} + \frac{y_1^2}{b^2}$

Generalising, the locus of (x_1, y_1) is $\frac{hx}{a^2} + \frac{ky}{b^2} = \frac{x^2}{a^2} + \frac{y^2}{b^2}$. **Ans.**

Example 19:

Find the locus of the intersection of tangents if the difference of these eccentric angles of 120°.

Solution:

$\phi_1 - \phi_2 = 120°$ (by hypothesis) ...(1)

The equation to tangent at point $(a \cos \phi_1, b \sin \phi_2)$ is

$(x/a) \cos \phi_1 + (y/b) \sin \phi_1 = 1$ ...(2)

The equation to tangent at point $(a \cos \phi_2, b \sin \phi_2)$ is

$$(x/a) \cos \phi_2 + (y/b) \sin \phi_2 = 1. \qquad \ldots(3)$$

On adding (2) and (3), we get

$$\frac{x}{a}(\cos \phi_1 + \cos \phi_2) + \frac{y}{b}(\sin \phi_1 + \sin \phi_2) = 2$$

$$\Rightarrow \quad \frac{x}{a}\left(2 \cos \frac{\phi_1 + \phi_2}{2} \cos \frac{\phi_1 + \phi_2}{2}\right) + \frac{y}{b}\left(2 \sin \frac{\phi_1 + \phi_2}{2} \cos \frac{\phi_1 + \phi_2}{2}\right) = 2$$

$$\Rightarrow \frac{x}{a} \cos \frac{\phi_1 + \phi_2}{2} + \frac{y}{b} \sin \frac{\phi_1 + \phi_2}{2} = \frac{1}{\cos \frac{\phi_1 + \phi_2}{2}} = \frac{1}{\cos \frac{120°}{2}} \quad \text{[by (1)]}$$

$$= \sec 60°.$$

$$\Rightarrow \quad \frac{x}{a} \cos \frac{\phi_1 + \phi_2}{2} + \frac{y}{b} \sin \frac{\phi_1 + \phi_2}{2} = 2 \qquad \ldots(4)$$

Again subtracting (2) from (3), we have as in the last question

$$\Rightarrow \quad \left(\frac{h^2}{a^2} + \frac{k^2}{b^2}\right)^2 = c^2\left(\frac{h^2}{a^4} + \frac{k^2}{b^4}\right).$$

Generalising, the locus of (h, k) is $\left(\frac{h^2}{a^2} + \frac{k^2}{b^2}\right)^2 = c^2\left(\frac{h^2}{a^4} + \frac{k^2}{b^4}\right)$. **Ans.**

Example 20:

Find the locus of the intersection of tangents if the sum of the ordinates of the points of contact be equal to b.

Solution:

The tangent at ϕ_1 is given as

$$(x/a) \cos \phi_1 + (y/b) \sin \phi_1 = 1,$$

and tangent at ϕ_2 is $(x/a) \cos \phi_2 + (y/b) \sin \phi_2 = 1$,

and $b \sin \phi_1 + b \sin \phi_2 = b$ (by hypothesis)

$\Rightarrow \quad \sin \phi_1 + \sin \phi_2 = 1$

$$\Rightarrow \quad 2 \sin \frac{\phi_1 + \phi_2}{2} \cos \frac{\phi_1 - \phi_2}{2} = 1. \qquad \ldots(3)$$

To find the locus of the required point, we have to eliminate ϕ_1 and ϕ_2 from (1), (2) and (3). From (1) and (2), on subtraction, we get

$$\frac{x}{a}(\cos\phi_1 - \cos\phi_2) + \frac{y}{b}(\sin\phi_1 - \sin\phi_2) = 0$$

$$\Rightarrow \quad \frac{2x}{a}\sin\frac{\phi_1+\phi_2}{2}\sin\frac{\phi_1-\phi_2}{2} + \frac{y}{b}\cos\frac{\phi_1+\phi_2}{2}\sin\frac{\phi_1-\phi_2}{2} = 0$$

$$\Rightarrow \quad \frac{x}{a}\sin\frac{\phi_1+\phi_2}{2} = \frac{y}{b}\cos\frac{\phi_1+\phi_2}{2}.$$

Putting the value of $\{\sin(\phi_1 + \phi_2)/2\}$ from (3), we get

$$\frac{x}{a}\cdot\frac{1}{2\cos\dfrac{\phi_1-\phi_2}{2}} = \frac{y}{b}\cos\frac{\phi_1+\phi_2}{2}$$

$$\Rightarrow 2\cos\frac{\phi_1+\phi_2}{2}\cos\frac{\phi_1-\phi_2}{2} = \frac{bx}{ay} \text{ or } \cos\phi_1 + \cos\phi_2 = \frac{bx}{ay} \quad \ldots(4)$$

Again by (1) and (2), on addition, we get

$$\frac{x}{a}(\cos\phi_1 + \cos\phi_2) + \frac{y}{b}(\sin\phi_1 + \sin\phi_2) = 2$$

$$\Rightarrow \quad \frac{x}{a}\cdot\frac{bx}{ay} + \frac{y}{b} = 2 \text{ from (4) and (3)}$$

$\Rightarrow$ $b^2x^2 + a^2y^2 = 2a^2by$ which is the required focus. **Ans.**

Example 21:

Prove that the locus of the pole, with respect to the ellipse, of any tangent to the auxilliary circle is the curve $\dfrac{x^2}{a^4} + \dfrac{y^2}{b^4} = \dfrac{1}{a^2}$.

Solution:

Any point on the auxiliary circle.

$x^2 + y^2 = a^2$ is $(a\cos\alpha, a\sin\alpha)$. The tangent to the circle at this point is

$$x\cos\alpha + y\sin\alpha = a. \quad \ldots(1)$$

Let (h, k) be the pole of this tangent with respect to the ellipse

$$\frac{x^2}{a^2} + \frac{y^2}{b^2} = 1. \quad \ldots(2)$$

As the equations (1) and (2) are identical, on comparing

$$\frac{\cos\alpha}{h/a^2} = \frac{\sin\alpha}{k/b^2} = \frac{a}{1}, \text{ where } \cos\alpha = \frac{h}{a} \text{ and } \sin\alpha = a\cdot\frac{k}{b^2}.$$

To eliminate α, on squaring the adding, we have

$$1 = \cos^2 \alpha + \sin^2 \alpha = \frac{h^2}{a^2} + \frac{a^2k^2}{b^4} \text{ or } \frac{1}{a^2} = \frac{h^2}{a^4} + \frac{k^2}{b^4}.$$

Generalising the locus of (h, k) is $\frac{x^2}{a^4} + \frac{y^2}{b^4} = \frac{1}{a^2}$. **Proved.**

Example 22:

Find the locus of the middle points of chords of the ellipse whose length is constant (= 2c).

Solution:

Suppose the co-ordinates of the middle point of the chord PQ of the ellipse $x^2/a^2 + y^2/b^2 = 1$, be (h, k). As the length of PQ is 2c so the co-ordinates of P and Q may be taken as $(h + c \cos \theta, k + c \sin \theta)$ and $(h - c \cos \theta, k - c \sin \theta)$ respectively. As these points lie on the ellipse, we must have

$$\frac{(h + c \cos \theta)^2}{a^2} + \frac{(k + c \sin \theta)^2}{b^2} = 1 \quad \text{...(1)}$$

and $$\frac{(h - c \cos \theta)^2}{a^2} + \frac{(k - c \sin \theta)^2}{b^2} = 1. \quad \text{...(2)}$$

Now, we have to eliminate θ from (1) and (2)

By (1) and (2) on adding, we have

$$(2/a^2)(h^2 + c^2 \cos^2 \theta) + (2/b^2)(k^2 + c^2 \sin^2 \theta) = 2$$

$$\Rightarrow \quad b^2 (h^2 + c^2 \cos^2 \theta) + a^2 (k^2 + c^2 \sin^2 \theta) = a^2b^2$$

$$\Rightarrow \quad b^2h^2 + a^2k^2 - a^2b^2 + c^2 (a^2 \sin^2 \theta + b^2 \cos^2 \theta) = 0. \quad \text{...(3)}$$

Again by (2) and (1) on subtraction, we have

$$\frac{4ch}{a^2} \cos \theta - \frac{4ck}{b^2} \sin \theta = 0$$

$$\Rightarrow \quad \frac{\sin \theta}{hb^2} = \frac{\cos \theta}{-ka^2} = \frac{1}{\sqrt{(h^2b^4 + k^2a^4)}}$$

Hence $\sin \theta = \frac{hb^2}{\sqrt{(h^2b^4 + k^2a^4)}}$ and $\cos q = \frac{-ka^2}{\sqrt{(h^2b^4 + k^2a^4)}}$

Putting the values of sin θ and cos θ in (3), we get

$$b^2h^2 + a^2k^2 - a^2b^2 + c^2\left(\frac{h^2b^4a^2}{(h^2b^4 + k^2a^4)} + \frac{k^2a^4b^2}{h^2b^4 + k^2a^4}\right) = 0$$

$\Rightarrow$ $(b^4h^2 + k^2a^4)(b^2h^2 + a^2k^2 - a^2b^2) + c^2a^2b^2(h^2b^2 + a^2k^2) = 0$

The Ellipse

Generalising, the locus of (h, k) is given as

$(b^4h^2 + a^4y^2)(b^2x^2 + a^2y^2 - a^2b^2) + c^2a^2b^2(x^2b^2 + a^2y^2) = 0$ **Ans.**

Example 23:

The polar of a point P with respect to an ellipse touches a fixed circle, whose centre is on the major axis and which passes through the centre of the ellipse. Show that the locus of P is a parabola, whose latus rectum is a third proportional to the diameter of the circle and the latus rectum of the ellipse.

Solution:

Suppose the equation to the ellipse be $\frac{x^2}{a^2} + \frac{y^2}{b^2} = 1$ ...(1)

and as the circle passes through (0, 0) and its centre is as x-axis, let centre by (c, 0), so the radius will be c and its equation will be

$(x - c)^2 + y^2 = c^2$. ...(2)

Now, if the co-ordinates cf P be (h, k); then it polar w.r.t. (1) is

$$\frac{xh}{a^2} + \frac{yk}{b^2} = 1.$$

If this tangent on the circle, then $\frac{b^2 ch - a^2b^2}{\sqrt{(b^4h^2 + a^4k^2)}} = \pm c$

(perpendicular from the centre must be equal to the radius)

$\Rightarrow$ $(b^2ch - a^2b^2)^2 - c^2(b^4h^2 + a^4k^2)$

$\Rightarrow$ $b^2c^2h^2 + a^4b^4 - 2cha^2b^4 = c^2b^4h^2 + c^2a^4k^2$

$\Rightarrow$ $c^2a^4k^2 = a^4b^4 - 2cha^2b^4$ or $c^2a^2k^2 = -2cb^4(h - a^2/2c)$.

Generalising, the locus of (h, k) is $y^2 = -\frac{2b^4}{ca^2}\left(x - \frac{a^2}{2c}\right)$.

This is a parabola having the latus rectum as $\frac{2b^4}{ca^2}$.

Now, if z be the third proportional of the diameter of the circle, and the latus rectum of the ellipse, then

$$2c : \frac{2b^2}{a} :: \frac{2b^2}{a} : z$$

(as the diameters of the circle and ellipse are respectively 2c and $2b^2/a$)

$$\Rightarrow \quad z = \frac{4b^2}{a^2 \cdot 2c} = \frac{2b^4}{a^2c} = \text{latus rectum at the parabola.} \quad \textbf{Proved.}$$

Example 24:

Find the locus of the middle points of chords of an ellipse the tangents at the ends of which intersect at right angles.

Solution:

The equation to the chord whose middle point is (h, k) will be

$$xh/a^2 + yk/b^2 = hk/a^2 + k^2/b^2 \quad \text{...(1)}$$

If (α, β) be the point of intersection of tangent at the extremities of the chord, then by hypothesis these intersect perpendicularly, hence the point (α, β) lies on the director circle whose equation is $x^2 + y^2 = a^2 + b^2$. ...(2)

The equation to the chord of contact of (a, b) w.r.t. ellipse

$$\frac{x^2}{a^2} + \frac{y^2}{b^2} = 1 \text{ is } \frac{\alpha x}{a^2} + \frac{\beta y}{b^2} = 1. \quad \text{...(3)}$$

Clearly the equations (1) and (3) represent the same line and hence are identical. Hence on comparison, as the last equation.

$$\alpha = \frac{a^2b^2h}{a^2h^2 + a^2k^2} \text{ and } \beta = \frac{a^2b^2k}{a^2h^2 + a^2k^2}$$

Since the point (α, β) lies on (2), hence we have

$$\frac{a^4b^4h^2 + a^4b^4k^2}{(b^2h^2 + a^2k^2)^2} = a^2 + b^2$$

$$\Rightarrow \quad a^4b^4 (h^2 + k^2) = (a^2 + b^2) (b^2h^2 + a^2k^2)^2$$

Generalising, the locus of (h, k) is

$$a^4b^4 (x^2 + y^2) = (a^2 + b^2) (b^2x^2 + a^2y^2)^2. \quad \textbf{Ans.}$$

Example 25:

A parallelogram circumscribes the ellipse and two of its opposite angular points lie on the straight lines $x^2 = h^2$; prove that the locus of the other two is the conic $x^2/a^2 + y^2/b^2 (1 - a^2/h^2) = 1$.

Solution:

Suppose the eccentric angles of the points of contact of the parallelogram which is curcumscribed on the ellipse $x^2/b^2 + y^2/b^2 = 1$ to be α, β, $\pi + \alpha$ and $\pi + \beta$ respectively. The tangent at 'α' will be $x/a \cos a + y/b. \sin \alpha = 1$ and the tangent at 'β' will be $x/a \cos \beta + y/\beta \sin \beta = 1$.

Solving these two, we get the co-ordinates of the point of intersection as

$$\left\{\frac{a \cos \frac{\alpha+\beta}{2}}{\cos \frac{\alpha-\beta}{2}} \cdot \frac{b \sin \frac{\alpha+\beta}{2}}{\cos \frac{\alpha-\beta}{2}}\right\} \quad \ldots(1)$$

As this point lies on $x^2 = h^2$ (by hypothesis).

$$\text{So} \qquad h = \frac{a \cos \frac{\alpha+\beta}{2}}{\cos \frac{\alpha-\beta}{2}}$$

$$\text{or} \qquad h \cos \frac{\alpha-\beta}{2} = a \cos \frac{\alpha+\beta}{2}. \quad \ldots(2)$$

Let (x, y) be the co-ordinates of the other angular points.

These may be obtained by putting $\alpha + \pi$ for α in (1).

$$\text{So} \qquad x = \frac{a \sin (\alpha+\beta)/2}{\sin (\alpha-\beta)/2} \quad \ldots(3)$$

$$\text{and} \qquad \frac{-b \cos \frac{\alpha+\beta}{2}}{\sin \frac{\alpha-\beta}{2}}$$

To obtain the locus of the point of intersection, we have to eliminate α and β from (2), (3) and (4).

By (4) and (2) on division, we have

$$\frac{y}{h} \tan \frac{\alpha-\beta}{2} = \frac{-b}{a}$$

$$\Rightarrow \qquad \cot \frac{\alpha-\beta}{2} = \frac{-ay}{bh}$$

Squaring and adding (3) and (4), we get

$$\frac{x^2}{a^2}+\frac{y^2}{b^2}=\frac{\sin^2\frac{\alpha+\beta}{2}+\cos^2\frac{\alpha+\beta}{2}}{\sin^2\frac{\alpha-\beta}{2}}=\operatorname{cosec}^2\frac{\alpha-\beta}{2}=1+\cot^2\frac{\alpha-\beta}{2}$$

and by (5), we get $\frac{x^2}{a^2}+\frac{y^2}{b^2}=1+\frac{a^2y^2}{a^2b^2}$ or $\frac{x^2}{a^2}+\frac{y^2}{b^2}\left(1-\frac{a^2}{h^2}\right)=1.$

Example 26:

Prove that the locus of the intersection of normals at the ends of chords, parallel to the tangent at the point whose eccentric angle is a, is the conic 2 (ax sin α + by cos α) (ax cos α + by sin α) = $(a^2-b^2)^2$ sin 2 α $\cos^2 2\alpha$.

If the chords be parallel to an equi-conjugate diameter, the locus is a diameter perpendicular to the other equi-conjugate.

Solution:

We know that the equation of tangent at point (a cos a, b sin a) of the ellipse $x^2/a^2+y^2/b^2$ is (x/a) cos a + (y/b) sin a = 1. ...(1)

The slope of the tangent is (–b/a) cot a. And the equation to the chord joining (a cos ϕ_1, b sin ϕ_1) and (a cos ϕ_2, b sin ϕ_2) is given as

$$\frac{x}{a}\cos\frac{\phi_1+\phi_2}{2}+\frac{y}{b}\sin\frac{\phi_1+\phi_2}{2}=\cos\frac{\phi_1-\phi_2}{2} \quad ...(2)$$

Its slope is $\frac{-b}{a}\cot\frac{\phi_1+\phi_2}{2}$.

If the lines (1) and (2) are parallel (by hypothesis).

Then $\frac{-b}{a}\cot\frac{\phi_1+\phi_2}{2}=\frac{-b}{a}\cot\alpha$, or $f_1+f_2=2a$. ...(3)

Suppose $\phi_1-\phi_2=2\theta$ (say), then $\phi_1=\alpha+\theta$, and $\theta_2=\alpha-\theta$.

So the co-ordinates of the ends will be

{a cos ($\alpha+\theta$), b sin ($\alpha+\theta$)} and {a cos ($\alpha-\theta$), b sin ($\alpha-\theta$)}

The normals at these points are respectively

ax sec ($\alpha+\theta$) – by cosec ($\alpha+\theta$) = a^2-b^2

$\Rightarrow$(ax sin α – by cos α) cos ($\alpha+\theta$) = (a^2-b^2) sin ($\alpha+\theta$) cos ($\alpha+\theta$)

$\Rightarrow$(ax sin α – by cos α) cos θ + (ax cos α + by sin α) sin θ

= 1/2 (a^2-b^2) sin ($2\alpha+2\theta$) ...(1)

and $\quad a \sec(\alpha - \theta) - by \operatorname{cosec}(\alpha - \theta) = a^2 - b^2$

$\Rightarrow ax \sin(\alpha - \theta) - by \cos(\alpha - \theta) = (a^2 - b^2)\sin(\alpha - \theta)\cos(\alpha - \theta)$

$\Rightarrow (ax \sin\alpha - by \cos\alpha)\cos\theta + (ax\cos\alpha + by\sin\alpha)\sin\theta$

$= 1/2\,(a^2 - b^2)\sin(2\alpha - 2\theta).$ ...(2)

Adding (1) and (2).

$2\,(ax\sin\alpha - by\cos\alpha)\cos\theta = (a^2 - b^2)\sin 2\alpha\cos 2\alpha\theta$ ...(3)

and subtracting (2) from (1),

$2\,(ax\cos\alpha + by\sin\alpha)\sin\theta = (a^2 - b^2)\cos 2\alpha\cos 2\theta$

$$\Rightarrow \quad \frac{(ax\cos\alpha + by\sin\alpha)}{(a^2 - b^2)\cos 2\alpha} = \cos\theta \qquad \text{...(4)}$$

and by (3),
$$\frac{\cos 2\theta}{\cos\theta} = \frac{2(ax\sin\alpha - by\cos\alpha)}{(a^2 - b^2)\sin 2\alpha} \qquad \text{...(5)}$$

As
$$\frac{\cos 2\theta}{\cos\theta} = \frac{2\cos^2\theta - 1}{\cos\theta} = 2\cos\theta - \frac{1}{\cos\theta}$$

Putting the values from (4),

$$\frac{2(ax\cos\alpha + by\sin\alpha)}{(a^2 - b^2)\cos 2\alpha} - \frac{(a^2 - b^2)\cos 2\alpha}{(ax\cos\alpha + by\sin\alpha)}$$

$$\frac{2(ax\cos\alpha + by\sin\alpha)^2 - (a^2 - b^2)\cos^2 2\alpha}{(a^2 - b^2)\cos 2\alpha\,(ax\cos\alpha + by\sin\alpha)}$$

Equating it to the R.H.S. of 5 and simplifying, we get

$2\,(ax\sin\alpha - by\cos\alpha)(ax\cos\alpha + by\sin\alpha)\cos 2\alpha$

$= [2\,(ax\cos\alpha + by\sin\alpha)^2 - (a^2 - b^2)\cos^2 2\alpha]\sin 2\alpha$

$\Rightarrow 2\,(ax\cos\alpha + by\sin\alpha)^2\,[(ax\sin\alpha - by\cos\alpha)\cos 2\alpha$

$- 2\,(ax\cos\alpha + by\sin\alpha)^2\sin 2a] = -(a^2 - b^2)\cos^2 2\alpha\sin 2\alpha$

$\Rightarrow -2\,(ax\cos\alpha + by\sin\alpha)\ ax\sin\alpha + by\cos\alpha)$

$= -(a^2 - b^2)^2\cos^2 2\alpha\sin 2\alpha.$

$\Rightarrow 2\,(ax\sin\alpha + by\cos\alpha)(ax\cos\alpha + by\sin\alpha)$

$= (a^2 - b^2)^2\sin 2\alpha\cos^2 2\alpha.$ **Proved.**

Again, if these chords be parallel to conjugate diameters, then $2\alpha = 90°$. Hence locus is $2\,(ax\sin 45° + by\cos 45°)(ax\cos 45° + by\sin 45°)$

$= (a^2 - b^2)^2\sin 90°\cos^2 90°$

or $(ax + by)(ax + by) = 0$

or $ax + by = 0.$

This is a diameter which is perpendicular to other equi-conjugate diameter given by the equation $ay - bx = 0$.

Example 27:

Chords at right angles are drawn through any point P of the ellipse, and the line joining their extremities meets the normal in the point Q. Prove that Q is the same for all such chords, its co-ordinates being

$$\frac{a^3e^2 \cos\alpha}{a^2 + b^2} \text{ and } \frac{-a^2 be^2 \sin\alpha}{a^2 + b^2}$$

Prove also that the major axis is the bisector of the angle PCQ, and that the locus of Q for different positions of P is the ellipse

$$\frac{x^2}{a^2} + \frac{y^2}{b^2} = \left(\frac{a^2 - b^2}{a^2 + b^2}\right)^2.$$

Solution:

Let the equation to the normal at any point $(a \cos\alpha, b \sin\alpha)$ of the ellipse $x^2/a^2 + y^2/b^2 = 1$ be $ax \sec a - by \operatorname{cosec}\alpha = a^2 - b^2$.

Transferring the origin to the point $(a \cos\alpha, b \sin\alpha)$ the equation to the ellipse becomes

$$\frac{1}{a^2}(x + a \cos a)^2 + \frac{1}{b^2}(y + b \sin a)^2 = 1$$

or $$\frac{x^2}{a^2} + \frac{y^2}{b^2} + \frac{2x\cos\alpha}{a} + \frac{2y\cos\alpha}{b} = 0 \qquad \text{...(1)}$$

and the equation to the normal will be

$$a(x + \cos\alpha)\sec\alpha - b(y + b\sin\alpha)\operatorname{cosec}\alpha = a^2 - b^2 \qquad \text{...(2)}$$

Let PL and PM be the chords of the ellipse; then $\angle LPM = 90°$ and let the equation to LM be

$$lx + my = 1. \qquad \text{...(3)}$$

Making the equation (1) homogeneous by (3), we have

$$\frac{x^2}{a^2} + \frac{y^2}{b^2} + \frac{2x\cos\alpha}{a}(lx + my) + \frac{2y\sin\alpha}{b}(lx + my) = 0$$

$$\because x^2\left(\frac{1}{a^2} + \frac{2l\cos\alpha}{a}\right) + y^2\left(\frac{1}{t^2} + \frac{2m\sin\alpha}{a}\right) + \dots = 0 \qquad \text{...(4)}$$

Equation (4) represents a pair of straight lines through the points of intersection of (3) and (1) and if, these lines are perpendicular, then coefficient of x^2 + co-efficient of $y^2 = 0$.

i.e.,

$$\frac{1}{a^2} + \frac{2l \cos \alpha}{a} + \frac{1}{b^2} + \frac{2m \sin \alpha}{a^2 + b^2} = 0$$

or

$$\frac{2}{ab}(bl \cos a + am \sin a) + \frac{a^2 + b^2}{a^2 b^2} = 0$$

$$\Rightarrow \quad bl \cos a + am \sin a = -\frac{a^2 + b^2}{2ab} \qquad \ldots(5)$$

Again solving (3) and (2), the co-ordinates of Q (x, y) will be given by

$$x = \frac{b \cos \alpha}{bl \cos \alpha + am \sin \alpha} = \frac{b \cos \alpha\, 2ba}{-(a^2 + b^2)} = \frac{-b^2 a}{a^2 + b^2} = \cos \alpha$$

and by (2) $\quad y = \frac{a}{b} x \tan a$ or $y = \frac{-2a^2 b}{-(a^2 + b^2)} .\sin \alpha.$

Changing to the original the co-ordinates of Q are

$$x = a \cos a + \frac{-2a^2 b \cos \alpha}{a^2 + b^2} = \frac{a(a^2 - b^2) \cos \alpha}{a^2 + b^2}$$

$$= \frac{a^3 e^2 \cos \alpha}{a^2 + b^2} \left(\text{as } \frac{a^2 - b^2}{a^2} = e^2 \right) \qquad \ldots(6)$$

and $\quad y = b \sin a + \frac{-2ab^2 \cos \alpha}{a^2 + b^2} = \frac{-(a^2 b - b^3) \sin \alpha}{a^2 + b^2}$

$$= \frac{-b(a^2 - b^2)}{a^2 + b^2} \sin \alpha = \frac{-a^2 b e^2 \sin \alpha}{a^2 + b^2} \qquad \ldots(7)$$

which is constant because a, b, e and α are constants.

To find the locus of Q, we have to eliminate α. By (6) and (7), we get

$$\frac{x^2}{a^2} + \frac{y^2}{b^2} = \frac{a^4 e^4 \cos^2 \alpha + a^4 e^4 \sin^2 \alpha}{(b^2 + b^2)^2} = \frac{a^4 e^4}{(a^2 + b^2)} = \frac{(a^2 e^2)^2}{(a^2 + b^2)^2}$$

and $\quad \frac{x^2}{a^2} + \frac{y^2}{b^2} = \frac{(a^2 - b^2)^2}{(a^2 + b^2)^2}$ is the required locus.

Again the equation to CP is $y = (b/a) x \tan \alpha$ and that of CQ is

$$y = (-b/a) x \tan \alpha.$$

Therefore, CP and CQ make equal angles with the axis and hence, the major axis bisects the angle PCQ.

Example 28:

Prove that the locus of the intersection of normals of the ends of conjugate diameters is the curve $2(a^2x^2 + b^2y^2)^3 = (a^2 - b^2)^2 (a^2x^2 - b^2y^2)^2$.

Solution:

Say P and Q be the ends of the conjugate diameters of the ellipse $x^2/a^2 + y^2/b^2 = 1$, and let the co-ordinates of P be $(a \cos \phi, b \sin \phi)$, then those of Q will be

$$[(a \cos (\phi + \pi/2), b \sin (\phi + \pi/2)] \text{ or } (-a \sin \phi, b \cos \phi).$$

Now, normal at P is $ax \sec \phi - by \operatorname{cosec} \phi = a^2 - b^2$ ...(1)

and Normal at Q is $ax \sec (\pi/2 + \phi) - by \operatorname{cosec} (\pi/2 + \phi) = a^2 - b^2$

or $-ax \operatorname{cosec} \phi - by \sec \phi = a^2 - b^2$ ...(2)

To find the required locus of intersection of (1) and (2). So subtracting (2) from (1), we get

$$(ax + by) \sec \phi - (by - ax) \operatorname{cosec} \phi = 0$$

or $$\frac{\cos \phi}{ax+by} = \frac{\sin \phi}{by-ax} = \frac{1}{\sqrt{[(ax+by)^2+(by-ax)^2]}} = \frac{1}{\sqrt{[2(a^2x^2+b^2y^2)]}}$$

whence $\sec \phi = \dfrac{\sqrt{[2(a^2x^2+b^2y^2)]}}{ax + by}$ and $\operatorname{cosec} \phi = \dfrac{\sqrt{2(a^2x^2+b^2y^2)}}{by - ax}$

Putting the value of $\sec \phi$ and $\operatorname{cosec} \phi$ in (1), we get

$$\sqrt{[2(a^2x^2+b^2y^2)]}\left(\frac{ax}{ax + by} - \frac{by}{by - ax}\right) = a^2 - b^2$$

On squaring and simplifying the required locus is

$$2(a^2x^2 + b^2y^2)^3 = (a^2 - b^2)^2 (a^2x^2 - b^2y^2)^2.$$ **Proved.**

Example 29:

Prove that the sum of the angles that the four normals drawn from any point to an ellipse make with the axis is equal to the sum of the angles that the two tangents from the same point make with the axis.

Solution:

The normal at any point (x_1, y_1) of the ellipse.

$$\frac{x^2}{a^2}+\frac{y^2}{b^2}=1 \text{ given by } \frac{x-x_1}{x_1/a^2}=\frac{y-y_1}{y_1/b^2}$$

or $$y=\frac{a^2y_1}{b^2x_1}.x=\frac{a^2y_1}{b^2x_1}.x_1+y_1 \quad ..(1)$$

The slope of the normal is $=\dfrac{a^2y_1}{b^2x_1}=m$ (say)

Again as (x_1, y_1) lies on the ellipse; so $(x_1{}^2/a^2)+(y_1{}^2/b^2)=1$

$\Rightarrow \quad b^2x_1{}^2+a^2y_1{}^2=a^2b^2.$

$$\therefore \quad b^2x_1^2+\frac{b^4m^2x_1^2}{a^2}=a^2b^2$$

(on putting the value of y_1 in terms of m)

$$\Rightarrow \quad x_1^2\left(a^2+b^2m^2\right)=a^4, \text{ hence } x_1=\frac{a^2}{\sqrt{(a^2+b^2m^2)}}.$$

$$\therefore \quad y_1=\frac{b^2mx_1}{a^2}=\frac{b^2m}{\sqrt{(a^2+b^2m^2)}}$$

Hence the equation to normal at (x_1, y_1), i.e., (1) becomes

$$y=mx-\frac{ma^2}{\sqrt{(a^2+b^2m^2)}}+\frac{b^2m}{\sqrt{(a^2+b^2m^2)}}$$

$$\Rightarrow \quad y=mx-\frac{\left(a^2-b^2\right)m}{\sqrt{(a^2+b^2m^2)}}$$

$$\Rightarrow \quad mx-y=\frac{\left(a^2-b^2\right)m}{\left(a^2+b^2m^2\right)}.$$

Let this normal pass through a fixed point (h, k) then

$$mh-k=\frac{\left(a^2-b^2\right)m}{\sqrt{(a^2+b^2m^2)}}$$

On squaring $(a^2+b^2m^2)(mh-k)^2=m^2(a^2-b^2)^2$

$$\Rightarrow \quad m^4b^2h^2-2m^3b^2hk+m^2\{(a^2h^2+b^2k^2)-(a^2-b^2)^2\}$$
$$-2a^2hkm+a^2k^2=0.$$

If the roots of this equation be m_1, m_2, m_3 and m_4; then

$$\Sigma m_1 = \frac{2k}{h} = S_1 \text{ (say)}, \ \Sigma m_1 m_2 = \frac{1}{b^2h^2}\left\{\left(a^2h^2 + b^2k^2\right)\right.$$

$$\left. - \left(a^2 - b^2\right)^2\right\} = S_2 \text{ (say)}.$$

$$\Sigma m_1 m_2 m_3 = \frac{2a^2k}{b^2h} = S_3 \text{ (say) and}$$

$$\Sigma m_1 m_2 m_3 m_4 = \frac{a^2k^2}{b^2h^2} = S_4 \text{ (say)}$$

Let θ_1, θ_2, θ_3 and θ_4 be the angles of inclination of normals with the axis corresponding to m_1, m_2, m_3 and m_4 respectively, then

$$\tan\theta_1 = m_1, \ \tan\theta_2 = m_2, \ \tan\theta_3 = m_3 \text{ and } \tan\theta_4 \ m_4$$

Now, $\tan(\theta_1 + \theta_2 + \theta_3 + \theta_4) = \dfrac{S_1 - S_3}{1 - S_2 + S_4}$

$$= \frac{\dfrac{2k}{h} - \dfrac{2a^2k}{b^2h}}{1 - \dfrac{1}{b^2h^2}\left\{a^2h^2 + b^2k^2 - \left(a^2 - b^2\right)^2\right\} \dfrac{a^2k^2}{b^2h^2}}$$

$$= \frac{2hk\left(b^2 - a^2\right)}{\left(b^2 - a^2\right)\left(h^2 - k^2 + b^2 - a^2\right)} = \frac{2hk}{\left(h^2 - k^2 + b^2 - a^2\right)} \qquad \text{...(2)}$$

Again let any tangent $y = m_1x + \sqrt{(a^2m_1^2 + b^2)}$ also pass through the point (h, k), then $k - m_1h = \sqrt{(a^2m_1^2 + b^2)}$. On squaring, $k^2 - 2m_1hk + m_1^2h^2 = a^2m_1^2 + b^2$ or $m_1^2(h^2 - a^2) - 2m_1hk + (k^2 - b^2) = 0$.

This equation will give us two roots of m_1 corresponding to the slopes of the tangents through (h, k). Let α and β be the angles of inclination of tangents with axis of x, then

$$\tan\alpha + \tan\beta = \frac{2hk}{h^2 - a^2} \text{ and } \tan a \tan b = \frac{k^2 - b^2}{h^2 - a^2}$$

$$\text{Hence } \tan(\alpha + \beta) = \frac{\tan\alpha + \tan\beta}{1 - \tan\alpha\tan\beta} = \frac{\dfrac{2hk}{h^2 - a^2}}{1 - \dfrac{k^2 - b^2}{h^2 - a^2}}$$

$$\frac{2hk}{h^2 - k^2 - a^2 + b^2} \qquad \text{...(3)}$$

$\therefore \tan(\Sigma\theta_1) = \tan(\alpha + \beta)$ or $\theta_1 + \theta_2 + \theta_3 + \theta_4 = a + b$. **Proved.**

Example 30(a):

If Y and Z be the feet of the perpendiculars from the foci upon the tangent at any point P of an ellipse, prove that the tangents of Y and Z to the auxilliary circle meet on the ordinate of P and that the locus of their point of intersection is another ellipse.

Solution:

Taking the co-ordinates of P on the ellipse $x^2/a^2 + y^2/b^2 = 1$ as $(a \cos \phi, b \sin \phi)$, the equation to the tangent at this point will be

$$(x/a) \cos \phi + (y/b) \sin \phi = 1. \quad \text{...(1)}$$

The equation to the auxilliary circle is $x^2 + y^2 = a^2$.

Let the feet of the perpendiculars from foci on the tangent at P be Y and Z. These must lie on the director circle. Let (h, k) be the co-ordinates of the point of intersection of tangents on Y and Z, then YZ will become the chord of contact of (h, k). So the equation of YZ is

$$xh + yk = a^2. \quad \text{...(2)}$$

(1) and (2) must be identical, so comparing, we get $h = a \cos \phi$, $k = (a^2/b) \sin \phi)$. Hence the ordinates of P lie on $x = a \cos \phi$.

To get the locus of (h, k), we have to eliminate ϕ.

As $\cos \phi = h/a$ and $\sin \phi = bk/a^2$, hence

On squaring and adding, we get $1 = h^2/a^2 + b^2k^2/a^4$

or $\quad a^2h^2 + b^2k^2 = a^4$.

Generalising the locus of (h, k) is $a^2x^2 + b^2y^2 = a^4$, which is an ellipse.

Example 30(b):

Prove that the directrixes of the two parabolas that can be drawn to have their foci at any given point P of the ellipse and to pass through its foci meet at an angle which is equal to twice the eccentric angle of P.

Solution:

Let the ellipse be $x^2/a^2 + y^2/b^2 = 1$ having foci as S (–ae, 0) and S' (ae, 0). Take any point P $(a \cos \phi, b \sin \phi)$ on it.

Then $PS = a + ae \cos \phi$ and $PS' = a - ae \cos \phi$.

Let the directrix of the parabola be $x \cos a + y \sin a = P$; then the point P $(a \cos \phi, b \sin \phi)$ is focus of the parabola and by hypothesis, it passes through S and S. Therefore, by the definition of the parabola, perpendicular from S on the directrix must be equal to SP.

or $\quad P + ae \cos a = a + ae \cos \phi.$ ...(1)

Again the perpendicular from S' on directrix = S'P

So $\quad P - ae \cos a = a - ae \cos \phi.$ ...(2)

From (2) and (1) on subtraction, we have $\cos a = \cos \phi$.

Hence $a = \phi$ or $-\phi$; therefore, the angle between the directrix is $\phi - (-\phi) = 2\phi$ which is double of the eccentric angle of P. **Proved.**

Example 30(c):

An ellipse is rotated through a right angle in its own plane about its centre, which is fixed; prove that the locus of the point of intersection of a tangent to the ellipse in its original position with the tangent at the same point of the curve in its new position is

$$(x^2 + y^2)(x^2 + y^2 = a^2 - b^2) = 2(a^2 - b^2)\, xy.$$

Solution:

We know that

$$x \cos \alpha + y \sin \alpha = \sqrt{(a^2 \cos^2 \alpha + b^2 \sin^2 \alpha)} \quad ...(1)$$

is always tangent on the ellipse $x^2/a^2 + y^2/b^2 = 1$.

The equation of the tangent to the ellipse when it has been rotated through 90° will be obtained by putting $\alpha + 90°$ for α in the L.H.S. of the original tangent, i.e., (1).

∴ Equation of tangent in the second position will be

$$x \cos (90° + \alpha) + y \sin (90° + \alpha) = \sqrt{(a^2 \cos^2 \alpha + b^2 \sin^2 \alpha)}$$

$$\Rightarrow \quad -x \sin \alpha + y \cos \alpha = \sqrt{(a^2 \cos^2 \alpha + b^2 \sin^2 \alpha)}. \quad ...(2)$$

From (2) and (1) on subtraction, we have

$$(y + x) \sin \alpha + (x - y) \cos \alpha = 0$$

$$\therefore \frac{\sin \alpha}{y - x} = \frac{\sin \alpha}{y + x} = \frac{1}{\sqrt{[(y - x)^2 + (y + x)^2]}} = \frac{1}{\sqrt{[2(y^2 + x^2)]}}.$$

Putting the values of $\sin \alpha$ and $\cos \alpha$ in (1), we have

$$x(y + x) + y(y - x) = \sqrt{[a^2 (y + x)^2 + b^2 (y - x)^2]}$$

$$\Rightarrow \quad x^2 + y^2 = \sqrt{[(a^2 + b^2)(x^2 + y^2) + 2xy(a^2 - b^2)]}$$

$$\Rightarrow \quad (x^2 + y^2)^2 = (a^2 + b^2)(x^2 + y^2) + 2xy(a^2 + b^2)$$

$$\Rightarrow \quad (x^2 + y^2)(x^2 + y^2 - a^2 - b^2) = 2xy(a^2 - b^2).$$ **Proved.**

Example 31:

Chords of the ellipse touch the parabola $ay^2 = -2b^2x$; prove that the locus of their poles is the parabola $ay^2 = 2b^2x$.

Solution:

Suppose the coordinates of the pole of any chord of the ellipse $x^2/a^2 + y^2/b^2 = 1$ be (h, k) then equation to the chord, i.e., the polar of (h, k) is

$$xh/a^2 + yk/b^2 = 1$$

or $$b^2xh + a^2yk = a^2b^2 \qquad \ldots(1)$$

On solving (1) with the given parabola $ay^2 = -2b^2x$, we have

$$\frac{-ay^2}{2} h + a^2 yk = a^2b^2$$

$$hy^2 - 2ayk + 2ab^2 = 0$$

If the line (1) is a tangent on the parabola then the roots of (2) must be coincident, i.e., the discriminant of (2) must be zero.

Hence $4a^2k^2 - 8ab^2h = 0$ or $ak^2 = 2b^2h$.

Generalising the locus of (h, k) is $ay^2 = 2b^2x$. **Proved.**

Example 32:

Triangles are formed by pairs of tangents drawn from any point on the ellipse $a^2x^2 + b^2y^2 = (a^2 + b^2)^2$ to the ellipse $x^2/a^2 + y^2/b^2 = 1$, and their chord of contact. Prove that the orthocentre of each such triangle lies on the ellipse.

Solution:

Take the point of contact to be (a cos α, b sin α) and (a cos β, b sin β) on the given ellipse $x^2/b^2 + y^2/b^2 = 1$. The equation to tangents at these points will be respectively:

$$(x/a) \cos \alpha + (y/b) \sin \alpha = 1$$

and $$(x/a) \cos \beta + (y/b) \sin \beta = 1.$$

Solving the two the points of intersection is

$$\left\{\frac{a \cos \frac{\alpha + \beta}{2}}{\cos \frac{\alpha - \beta}{2}} \cdot \frac{b \sin \frac{\alpha + \beta}{2}}{\cos \frac{\alpha - \beta}{2}}\right\}$$

∴ This point lies on the ellipse $a^2x^2 + y^2b^2 = (a^2 + b^2)^2$.

$$\therefore \quad \frac{a^2 a^2 \cos^2 \dfrac{\alpha+\beta}{2}}{\cos^2 \dfrac{\alpha-\beta}{2}} + \frac{b^2 b^2 \sin^2 \dfrac{\alpha+\beta}{2}}{\cos^2 \dfrac{\alpha-\beta}{2}} = \left(a^2+b^2\right)^2$$

$$\text{or } a^4 \cos^2 \frac{\alpha+\beta}{2} + b^4 \sin^2 \frac{\alpha+\beta}{2} = (a^2+b^2)\cos^2 \frac{\alpha-\beta}{2} \quad \ldots(1)$$

The equation of perpendicular line from $(a \cos \beta, b \sin \beta)$ on the tangent $(x/a) \cos \alpha + (y/b) \sin \alpha = 1$, will be

$$a \sin \alpha(x - a \cos \beta) - b \cos \alpha\,(y - b \sin \beta) = 0. \quad \ldots(2)$$

Similarly the equations to other perpendicular will be

$$a \sin \beta\,(x - a \cos \alpha) - b \sin \beta\,(y - b \sin \alpha) = 0. \quad \ldots(3)$$

Solving (2) and (3), we get

$$x = \frac{a^2}{a^2+b^2} \frac{\cos \dfrac{\alpha+\beta}{2}}{\cos \dfrac{\alpha-\beta}{2}} \text{ and } y = \frac{b^2}{a^2+b^2} \frac{\sin \dfrac{\alpha+\beta}{2}}{\sin \dfrac{\alpha-\beta}{2}}$$

If the orthocentre of this triangle lies on the ellipse, $x^2/a^2 + y^2/b^2 = 1$, then we must have

$$x^2 \cos^2 \frac{\alpha+\beta}{2} + b^4 \sin^2 \frac{\alpha+\beta}{2} = (a^2+b^2)^2 \cos^2 \frac{\alpha-\beta}{2}.$$

which by (1) is true.

Therefore the orthocentre of the triangle lies on the ellipse. **Proved.**

Example 33(a):

Given the base of a triangle and the sum of its sides, prove that the locus of the centre of its in circle is an ellipse.

Solution:

Let ABK be any triangle.

Taking BC as axis of x and a perpendicular line through O the mid-point of BC as the y-axis.

Suppose the co-ordinates of the in-centre I of the circle be (x, y). Draw perpendicular IN on BC.

$$\therefore \quad x = ON = BN - BO = (s-b) - \frac{a}{2} = \frac{a+b+c}{2} - b - \frac{a}{2}$$

where $s = \frac{a + b + c}{2}$

Hence $x = \frac{c - b}{2}$

Again the area of the triangle ABC = $\sqrt{\{s(s-a)(s-b)(s-c)}$

$$= \sqrt{\left\{s(s-a)\frac{a+c-b}{2}\cdot\frac{(a+b-c)}{2}\right\}}$$

$$= \frac{1}{2}\sqrt{\left[s(s-a)\left\{a^2 - (c-b)^2\right\}\right]}$$

$$= \frac{1}{2}\sqrt{\left\{s(s-a)\left(a^2 - 4x^2\right)\right\}} \text{ from (1).}$$

If the radius of the in-circle be r, clearly r is equal to the ordinate of the centre.

So

$$r = y\frac{\Delta}{s} = \frac{1}{2s}\sqrt{\left\{s(s-a)\left(a^2 - 4x^2\right)\right\}}$$

Putting $l = \sqrt{(a^2 - 4x^2)}$ or $y^2 = \lambda^2(a^2 - 4x^2)$ or $4\lambda^2x^2 + y^2 = a^2\lambda^2$, which is the required locus and is clearly an ellipse.

Example 33(b):

Find the equation of the ellipse referred to its centre whose latus rectum is 5 and whose eccentricity is 2/3.

Solution:

Let the equation to the ellipse be $\frac{x^2}{a^2} + \frac{y^2}{b^2} = 1$ ...(1)

Latus rectum = $\frac{2b^2}{a} = 5$ and $e = \frac{2}{3}$ (by hypothesis).

As $b^2 = a^2(1 - e^2)$

So $e^2 = 1 - \frac{b^2}{a^2} = \frac{4}{9}$, whence $\frac{b^2}{a^2} = \frac{4}{9}$ ...(2)

Dividing $\frac{2b^2}{a} = 5$ by (2), we get $a = \frac{5}{9}$

Putting in (2), $b^2 \frac{5a}{2} = \frac{5}{2} \times \frac{9}{2} = \frac{45}{4}$

Putting the value of a^2 and b^2 in (1), we get

$$\frac{4x^2}{81} + \frac{4y^2}{45} = 1$$

or $20x^2 + 36y^2 = 405.$ **Ans.**

Example 34(a):

Find the equation of the ellipse referred to its centre whose minor axis is equal to the distance between the foci and whose latus rectum is 10.

Solution:

Let the ellipse be $\frac{x^2}{a^2} + \frac{y^2}{b^2} = 1.$ ...(1)

Minor axis = 2b = 2ae (by hypothesis). or b = ae, ...(2)

and latus rectum = $\frac{2b^2}{a} = 10$ (by hypothesis). ...(3)

and we know that $b^2 = a^2 (1 - e^2)$ ...(4)

Solving (2), (3) and (4), we get $a^2 = 100$ and $b^2 = 50$.

Putting in (1), the equation to the ellipse is $\frac{x^2}{100} + \frac{y^2}{50} = 1.$

or $x^2 + 2y^2 = 100$ **Ans.**

Example 34(b):

Find the eccentricity of an ellipse, if its latus rectum be equal to one half its minor axis.

Solution:

If the equation to the ellipse be $\frac{x^2}{b^2} + \frac{y^2}{a^2} = 1.$ then latus rectum $= \frac{2b^2}{a} = b$ (by hypothesis) or $\frac{b}{a} = \frac{1}{2}$

So $e = \sqrt{\left(1 - \frac{1}{4}\right)} = \sqrt{\frac{3}{4}} = \frac{\sqrt{3}}{2}.$ **Ans.**

Example 35:

Find the equation to the ellipse, whose focus is the point (–1, 1), whose directrix is the straight line $x - y + 3 = 0$, and whose eccentricity is 1/2.

Solution:

If P (x, y) be any point on the ellipse, S be its focus, and PN be the perpendicular from P on directrix, then by definition of the ellipse, $PS^2 = e^2 PN^2$, hence

$$(x+1)^2 + (y-1)^2 = \frac{1}{4}\left(\frac{x-y+3}{\sqrt{2}}\right)^2 = \left(\frac{x-y+3}{8}\right)^2;$$

[As focus is (–1, 1) and directrix is $x - y + 3 = 0$]

$$\Rightarrow \quad 8(x^2 + y^2 + 2x - 2y + 2) = x^2 + y^2 + 9 = 2xy + 6x - 6y$$

$$\Rightarrow \quad 7x^2 + 2xy + 7y^2 + 10x - 10 + 7 = 0.$$ **Ans.**

Example 36:

Is the point (4, –3) within or without the ellipse $5x^2 + 7xy^2 = 11$?

Solution:

As the equation of the ellipse is

$$5x^2 + 7y^2 - 11 = 0 \qquad ...(1)$$

and the point is (4, –3).

Putting the co-ordinates (4, –3) in (1), we have

L.H.S. = 80 + 63 – 11 = 132 which is positive.

Hence the point (4, –3) is out of the ellipse. **Ans.**

Example 37(a):

Find the locus of the middle points of chords on ellipse which are drawn through the positive end of the minor axis.

Solution:

Let $(a \cos \phi, b \sin \phi_2)$ be the co-ordinates of the other extremity of the chord of ellipse $\frac{x^2}{2} + \frac{y}{b^2} = 1$. The positive end of the minor axis is clearly (0, b). Let (x, y) be the middle point of chord; then

$$x = \frac{a \cos \phi + 0}{2} \text{ and } y = \frac{b + b \text{ sub } \pi}{2}$$

Hence $\quad \cos \phi = \frac{2x}{a} \qquad ...(1)$

and $$\sin\phi = \frac{2y-b}{b} \qquad ...(2)$$

Squaring and adding (1) and (2), we get

$$1 = \frac{4x^2}{a^2} + \left(\frac{2y-b}{b}\right)^2 \quad \text{or} \quad \frac{4x^2}{a^2} + \frac{4y^2-4yb+b^2}{b^2} - 1 = 0$$

Simplifying, the required locus of (x, y) is $\frac{x^2}{a^2}+\frac{y^2}{b^2} = \frac{y}{b}$. **Ans.**

Example 37(b):

Prove that the area of the triangle by three points on an ellipse, whose eccentric angles are θ, φ and ψ is

$$2ab \sin\frac{\phi-\psi}{2}\sin\frac{\psi-\theta}{2}\sin\frac{\theta-\phi}{2}.$$

Solution:

(a) the co-ordinates of the given points on the ellipse

$$\frac{x^2}{a^2}+\frac{y^2}{b^2} = 1$$

will be (a cos θ, b sin θ), (a cos φ, b sin φ) and (a cos ψ, b sin ψ).

Area of the Δ formed by these points.

$$= \frac{1}{2}(a\cos\theta,\ b\sin\theta),\ (a\cos\phi,\ b\sin\phi) \text{ and } (a\cos\psi,\ b\sin\psi).$$

$$-ab\sin\phi\cos\psi + ab\sin\theta\cos\psi - ab\cos\theta\sin\psi]$$

$$= \frac{1}{2}ab\ [\sin(\phi-\theta) + \sin(\psi-\phi) + \sin(\theta-\psi)]$$

$$= \frac{1}{2}ab\left(2\sin\frac{\phi-\theta}{2}\cos\frac{\phi-\theta}{2} + 2\frac{\psi-\theta}{2}\cos\frac{\psi-\theta}{2}\right.$$

$$\left. + 2\sin\frac{\theta-\theta}{2}\cos\frac{\phi-\psi}{2}\right)$$

$$= ab\sin\frac{\theta-\phi}{2}\left(\cos\frac{\theta-\phi-2\psi}{2} / \cos\frac{\phi-\theta}{2}\right) = ab\sin\frac{\theta-\phi}{2}$$

$$\sin\frac{\phi-\psi}{2}\sin\frac{\psi-\theta}{2}$$ **Proved.**

Example 38(a):

Find the equation of the ellipse referred to its centre whose foci are the points (4, 0) and (–4, 0) and whose eccentricity is 1/3.

Solution:

Let the equation to the ellipse be $\frac{x^2}{a^2} + \frac{y^2}{b^2} = 1$ ...(1)

Distance between the foci = 2ae = 4 + 4 = 8 as foci are (± 4, 0) ...(2)

Putting the value of $e = \frac{1}{3}$ (given) in (2), $a = \frac{4}{e} = 4 \times 3 = 12.$

Again $b^2 = a^2 (1 - e^2) = 144 \left(1 - \frac{1}{9}\right) = 144 \times \frac{8}{9} = 128.$

Putting the values in (1), we have $\frac{x^2}{144} + \frac{y^2}{128} = 1.$

$8x^2 + 9y^2 = 1, 152.$ **Ans.**

Example 38(b):

Any point P of an ellipse is joined to the extremities of the major axis; prove that the portion of a directrix intercepted by them subtends a right angle at the corresponding focus.

Solution:

Let the ellipse be

$$\frac{x^2}{a^2} + \frac{y^2}{b^2} = 1$$

and P (a cos ϕ, b sin ϕ) by a point on it.

The equation to PA is given as

$$y - 0 = \frac{b \sin \phi - 0}{a \cos \phi} (x - a)$$

It meets the directrix x = a/e in N whose co-ordinates are

$$\left[\frac{a}{e}, b \sin \phi \frac{(1 - e)}{e (\cos \phi - 1)}\right]$$

Equation to PA' is $y = \frac{b \sin \phi}{a \cos \phi + a} (x + a).$

It meets the directrix x = a/e in M whose co-ordinates are

$$\left[\frac{a}{e}, b \sin \phi \frac{1 + e}{e (\cos \phi + 1)}\right]$$

Now product of gradients of SN and SM we have

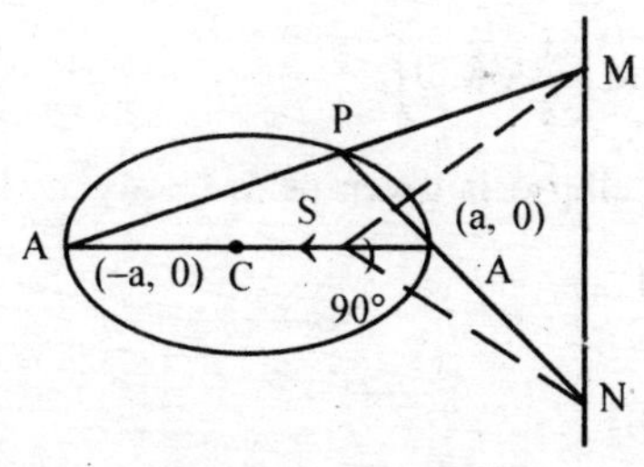

$$\left(\frac{\dfrac{b + \sin\theta\,(1 - e)}{e + (\cos\theta - 1)}}{a/e - ae}\right)\left(\frac{\dfrac{b + \sin\theta\,(1 + e)}{e + (\cos\theta + 1)}}{a/e - ae}\right)$$

$$= \left(\frac{\dfrac{b\sin\theta\,(1 - e)}{e\,(\cos\theta - 1)}}{a\,(1 - e^2)}\right)\left(\frac{\dfrac{b\sin\theta\,(1 + e)}{e\,(\cos\theta + 1)}.e}{a\,(1 - e^2)}\right)$$

$$= \frac{b^2\ 1 - e^2\ \sin^2\theta}{(\cos^2\theta - 1).\,a^2\,(1 - e^2)} = \frac{a^2\,(1 - e^2)\,(1 - e^2)\sin^2\theta}{-a^2(1 - e^2)\,(1 - \cos^2\theta)} = -1$$

i.e. SN is perpendicular to SM or MN subtends a right angle at S. **Proved.**

Example 39:

Find the latus rectum, the eccentricity, and the coordinates and the foci, of the ellipses (1) $x^2 + 3y^2 = a^2$, (2) $5x^2 + 4y^2 = 1$, and (3) $9x^2 + 5y^2 - 30y = 0$.

Solution:

(i) The equation to the ellipse is given as $x^2 + 3y^2 = a^2$.

$$\Rightarrow \quad \frac{x^2}{2} + \frac{y^2}{a^2} = 1 \qquad ...(1)$$

Hence $b^2 = a^2/3$

$$\therefore \text{ Latus rectum} = \frac{2b^2}{a} + \frac{a^2/3}{a} = \frac{2a}{3}.$$

$$\text{Eccentricity} = e = \sqrt{\left(\frac{a^2 - b^2}{a^2}\right)} = \sqrt{\left(\frac{a^2 - a^2/3}{a^2}\right)} = \frac{\sqrt{2}}{3} = \frac{1}{3}\sqrt{6}.$$

As the co-ordinates of foci are (ae, 0), so the foci are

$$\left(\frac{a}{3}\sqrt{6}, 0\right) \text{ and } \left(\frac{a}{3}\sqrt{6}, 0\right).$$ **Ans.**

(ii) The equation to the ellipse is given as $5x^2 + 4y^2 = 1$

or $$\frac{x^2}{1/5} + \frac{y^2}{1/4} = 1$$

Hence $a^2 = \frac{1}{5}$ and $b^2 = \frac{1}{4}$

Here as b > a, so

$$\text{Latus rectum} = \frac{2a^2}{b} = \frac{2\frac{1}{5}}{\frac{1}{2}} = \frac{4}{5}$$

$$\therefore\ e = \sqrt{\left(\frac{b^2 - a^2}{b^2}\right)} = \sqrt{\left(\frac{\frac{1}{4} - \frac{1}{5}}{\frac{1}{4}}\right)} = \frac{1}{\sqrt{5}} = \frac{\sqrt{5}}{5}$$

foci are $(0 \pm be)$ hence $\left(0 - \frac{1}{10}\sqrt{5}\right)$ **Ans.**

(iii) The equation to the ellipse is give as

$$9x^2 + 5y^2 - 30y = 0 \text{ or } 9x^2 + 5(y^2 - 6y + 9 - 9) = 0$$

$\Rightarrow$ $9x^2 + 5$ and $b^2 = 9$, so again b > a

$$e = \sqrt{\left(\frac{b^2 - a^2}{b^2}\right)} = \sqrt{\left(\frac{9 - 5}{9}\right)} = \frac{2}{3}$$

and

$$\text{So the latus rectum} = \frac{2a^2}{b} = \frac{2.5}{3} = \frac{10}{3}$$

Changing the origin of (1) to (0, 3) the foci will be $\left(0, \pm 3\frac{2}{3}\right)$, hence according to the original origin, foci are

(0, ±2 +3). i.e., (0, 5) and (0, 1) **Ans.**

Example 40:

Prove that the locus of the intersection of AP with the straight line through A' perpendicular to A'P is a straight line which is perpendicular to the major axis.

Solution:

Let the equation to the ellipse be $\frac{x^2}{a^2} + \frac{y^2}{b^2}$ ... be (a, 0) and (–a, 0). If

The co-ordinates of A' and A wil... ellipse, then the equation to AP

$P \equiv (a \cos \phi, b \sin \phi)$ be any ...

will be

$$\ldots = \frac{\ldots \sin \phi - 0}{a \cos \phi + a}(x + a)$$

$$\Rightarrow \quad y = \frac{b}{a} \frac{2 \sin \phi/2 \cos \phi/2}{2 \cos^2 \phi/2}(x + a)$$

$$\Rightarrow \quad y = \frac{b}{a} \tan \frac{\phi}{2}(x + a) \qquad \ldots(1)$$

Similarly equation of A'P is $y - 0 = \frac{b \sin \phi - 0}{a \cos \phi - a}(x + a)$

$$\Rightarrow \quad y = \frac{-b}{a} \cot\left(\frac{\phi}{2}\right)(x - a) \quad \left(\because \cos \phi - 1 = 2 \sin^2 \frac{\phi}{2}\right)$$

Equation of line through A' and perpendicular to A'P

$$y = (a/b) \tan (\phi/2)(x - a) \qquad \ldots(2)$$

The required locus of the point of intersection of (1) and (2) will be obtained by eliminating the variable ϕ between the two.

So dividing (1) by (2), we have

$$1 = \frac{b^2 x + a}{a^2 x - a} \quad \text{or } x = a \ (a^2 = b^2).$$

It is clearly a straight line which is perpendicular to the x-axis, i.e., the major axis.

Example 41:

Prove that the sum of the squares of the reciprocals of two perpendicular diameters of an ellipse is constant.

Solution:

Let the equation of the ellipse be $x^2/a^2 + y^2/b^2 = 1$.

Putting $x = r \cos \theta$ and $y = r \sin \theta$; we have the polar equation as

$$\frac{r^2 \cos^2 \theta}{a^2} + \frac{r^2 \sin^2 \theta}{b^2} = 1 \text{ or } \frac{\cos^2 \theta}{a^2} + \frac{\sin^2 \theta}{b^2} = \frac{1}{r^2}$$

angi here r is the radius vector. And if r and r_1 be the perpendicular radii ...ctively then the diameters will be 2r and $2r_1$ and the corresponding ...nd $(90° + \theta)$. So we have

$$\frac{1}{4r^2} + \frac{1}{4r_1^2} = \ldots 4 \left(\ldots \right)$$

$$= \frac{1}{4}\left[\left(\frac{\cos^2\theta}{a^2} + \frac{\sin^2\theta}{r^2}\right) + \left(\frac{\cos^2 (\ldots}{a^2} \ldots \frac{\sin^2(90° + \theta)}{\ldots}\right)\right]$$

$$= \frac{1}{4}\left[\frac{\cos^2\theta}{a^2} + \frac{\sin^2\theta}{b^2} + \frac{\sin^2\theta}{a^2} + \frac{\cos^2\theta}{b^2}\right]$$

$$= \frac{1}{4}\left[\frac{1}{a^2}(\cos^2\theta + \sin^2\theta) + \frac{1}{b^2}(\cos^2\theta + \sin^2\theta)\right]$$

$$= \frac{1}{4}\left(\frac{1}{a^2} + \frac{1}{b^2}\right)$$ which is a constant quantity. **Proved.**

Example 42(a):

Show that the perpendiculars from the centre upon all chords, which join the ends of perpendicular diameters, are of constant length.

Solution:

Let OL and OM be the semi-perpendicular diameter of the ellipse $x^2/a^2 = y^2/b^2 = 1$.

Suppose the equation of LM is $x \cos\alpha + y \sin\alpha = P$. ...(2)
so that the perpendicular from centre O upon LM is p.

To find the combined equation of OL and OM, we have to make (1) homogeneous with the help of (2); so we get

$$\frac{x^2}{a^2} + \frac{y^2}{b^2} = \left(\frac{x\cos\alpha + y\sin\alpha}{P}\right)^2$$

$$\Rightarrow x^2\left(\frac{1}{a^2} - \frac{\cos^2\alpha}{p^2}\right) - \frac{2xy\sin\alpha + \cos\alpha}{P^2} + y^2\left(\frac{1}{b^2} - \frac{\sin^2\alpha}{p^2}\right) = 0$$

As the lines OL and OM are mutually perpendicular, the coefficient of x^2 + coefficient of y^2 must be equal to zero.

So $$\frac{1}{a^2} - \frac{\cos^2\alpha}{p^2} + \frac{1}{b^2} - \frac{\sin^1\alpha}{p^2} = 0$$

$$\Rightarrow \quad \frac{1}{a^2}+\frac{1}{b^2}-\frac{\sin^1\alpha+\cos^2\alpha}{p^2}=\frac{1}{p^2}$$

$$\Rightarrow \quad p = ab/\sqrt{(a^2+b^2)}.$$

Hence the length of perpendicular p from O (0, 0) on the line is constant.

Example 42(b):

Find the lengths of the focal radii drawn to the point $(4\sqrt{3}, 5)$ of the ellipse $25x^2 + 16y^2 = 1600$.

Solution:

The equation to the ellipse is given as $25x^2 + 16y^2 = 1600$.

$$\Rightarrow \quad \frac{x^2}{64}+\frac{y^2}{100}=1.$$

Hence $a^2 = 64$ and $b^2 = 100$.

As $b > a$, we have

$$e = \sqrt{\left(\frac{b^2-a^1}{b^2}\right)} + \sqrt{\left(\frac{100-64}{100}\right)} = \frac{3}{5}$$

and be = 10 × 3/5 = 6. Hence the co-ordinates of foci (S and S'), {i.e., (0, be) and (0, –be) are (0, 6). The co-ordinates of point P are $(4\sqrt{3}, 5)$.

Hence

$$PS = \sqrt{[(4\sqrt{3})^2 + (5-6)^2]} = \sqrt{49} = 7$$

and

$$PS' = \sqrt{[(4\sqrt{3})^2 + (5+6)^2]}$$
$$= \sqrt{(48+121)} = \sqrt{(169)} = 13.$$

Again the equation of PS is $y - 6 = \dfrac{5-6}{4\sqrt{3}-0}(x-0)$

$$\Rightarrow \quad x + 4y\sqrt{3} - 24\sqrt{3} = 0$$

and the equation of PS' is $y + 6 = \dfrac{5-6}{4\sqrt{3}-0}(x-0)$

$$\Rightarrow \quad 11x - 4y\sqrt{3} - 24\sqrt{3} = 0.$$ **Ans.**

Example 42(c):

Prove that the perpendicular from the focus upon any tangent and the line joining the centre to the point of contact meet on the corresponding directrix.

Solution:

The equation of the tangent at point P (a cos ϕ, b sin ϕ) of the ellipse

$\frac{x^2}{a^2} + \frac{y^2}{b^2} = 1$ will be

$$\frac{x}{a}\cos\phi + \frac{y}{b}\sin\phi = 1$$

or $$bx\cos\phi + ay\sin\phi = ab. \qquad ...(1)$$

Equation tot he line SY through the focus S (ae, 0) and perpendicular to (1) is

$$a\sin\phi\,(x - ae) - b\cos\phi.\ y = 0 \qquad ...(2)$$

Also the equation to CP is

$$y = \frac{b\sin\phi}{a\cos\phi}x \qquad ...(3)$$

where C is the cenţre (0, 0).

Solving (2) and (3), we get x = a/e.

This is the equation to the directrix. Hence SY and CP meet on the directrix.

Example 43(a):

Prove that the straight lines, joining each focus to the foot of the perpendicular from the other focus upon the tangent at any point P, meet on the normal PG and bisect it.

Solution:

Suppose YPN' be any tangent at a point P of the ellipse. Let its foci be S and S'. Drop perpendiculars SY and S'Y' on the tangent and say SY and S'Y' produced meet in L and PG is the normal.

In the ΔSS'L, SY' will be the median. As PG is parallel to LS' (both being perpendicular to YPN'), therefore SY' bisects PG.

Similarly, we can proved that S'Y also bisects PG. This means SY' meet on the normal PG and bisect it.

Example 43(b):

Prove that the straight line lx + my = n is a normal to the ellipse, if

$$\frac{a^2}{l^2} + \frac{b^2}{m^2} = \frac{(a^2 - b^2)}{n^2}.$$

Solution:

The normal at any point P (a cos ϕ, b sin ϕ) of the ellipse $x^2/a^2 + y^2/b^2 = 1$ is given by

$$\frac{ax}{\cos\phi} - \frac{by}{\sin\phi} = a^2 - b^2. \quad ...(1)$$

If the given line $\quad l\,x + my = n \quad ...(2)$

is also a normal, then the lines (1) and (2) are identical. Therefore comparing the coefficients, we get

$$\frac{l\cos\phi}{a} = \frac{m\sin\phi}{-b} = \frac{n}{a^2 - b^2}.$$

$$\text{So } \cos\phi = \frac{an}{l(a^2 - b^2)} \text{ and } \sin\phi = \frac{-bn}{m(a^2 - b^2)},$$

To eliminate ϕ, we have

$$\cos^2\phi + \sin^2\phi = \frac{a^2 n^2}{l^2 (a^2 - b^2)} + \frac{b^2 n^2}{m^2 (a^2 - b^2)},$$

$$\Rightarrow \quad I = \frac{a^2 n^2}{l^2 (a^2 - b^2)} + \frac{b^2 n^2}{m^2 (a^2 - b^2)} \text{ or } \frac{a^2}{l^2} + \frac{b^2}{m^2} = \frac{(a^2 - b^2)^2}{n^2}.$$

Example 43(c):

PM and PN are perpendiculars upon the axes from any point P on the ellipse. Prove that MN is always normal to a fixed concentric ellipse.

Solution:

If the co-ordinates P on the ellipse be (a cos ϕ, b sin ϕ), as PM and PN are perpendiculars on x-axis and y-axis respectively, the co-ordinates of M and N will be (a cos ϕ, 0) and (0, b sin ϕ) respectively.

$$\text{Now equation to MN is } y - 0 \; \frac{b\sin\phi - 0}{0 - a\cos\phi}(x - a\cos\phi)$$

$$\Rightarrow \quad x/(a\cos\phi) + y/(b\sin\phi) = 1$$

$$\Rightarrow \quad (x/a\sec\phi) + (y/b)\sin\phi) = 1 \quad ...(1)$$

Now equation to the normal at point (a cos ϕ, b sin ϕ) with respect to any other concentric ellipse

$$x^2/A^2 + y^2/B^2 = 1 \text{ is}$$

$$Ax\sec\phi - By\,\text{cosec}\,\phi = A^2 = B^2. \quad ...(2)$$

As (1) and (2) are similar, on comparing them, we have

$$\frac{A}{l/a} = \frac{-B}{l/b} A^2 - B^2$$

$\Rightarrow$ $Aa = -Bb = A^2 = B^2.$ ...(3)

Solving the two equations, given by relation (3), we get

$$B = \frac{a^2 b}{a^2 - b^2} \text{ and } A = \frac{ab^2}{(a^2 - b^2)}$$

So the line (1), i.e., (x/a) sec ϕ + (y/b) cosec ϕ = 1 is a normal to the fixed ellipse

$\frac{x^2}{A^2} + \frac{y^2}{B^2} = 1$, where $A = \frac{-ab^2}{a^2 - b^2}$ and $B = \frac{a^2 b}{a^2 - b^2}$, both of them being constant.

Example 44:

Prove that the extremities of the latus recta of all ellipses, having a given major 2a, lie on the parabola x2 = –a (y – a), or on the parabola x^2 = a (a + a).

Solution:

Let LSL' be the latus rectum, c be the centre of the ellipse, and the co-ordinates of L be (x, y), then

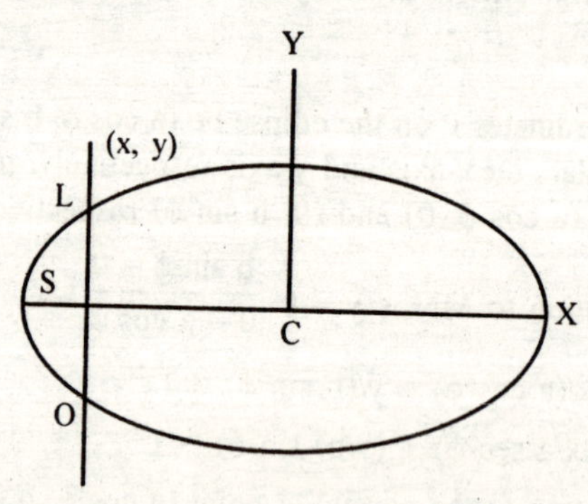

$x = CS = ae$ ...(1)

and $y = SL = \frac{b^2}{a} = \frac{a^2(1 - e^2)}{a}$

$\Rightarrow$ $y = a(1 - e^2)$ ...(2)

Eliminating the variable e from [1 and 2], we get the locus of L. Hence putting the value of e from (1) in (2), we get

$$y = a\,(1 - x^2/a^2)$$

or $$x^2 = -a\,(y - a)$$

which is clearly a parabola.

Similarly, we can show that the locus of L is $x^2 = (y + a)$ which is again a parabola.

Example 45:

If $\theta_1, \theta_2, \theta_3$ and θ_4 be the eccentric angles of the four points of intersection of the ellipse and any circle, prove that $\theta_1 + \theta_2 + \theta_3 + \theta_4$ is an even multiple of π radians.

Solution:

Suppose the equation of teh ellipse is

$$x^2/a^2 + y^2/b^2 = 1 \qquad ...(1)$$

and $y = b \sin \theta$ must satisfy the equation to the circle so putting in (1),

$$a^2 \cos^2 \theta + b^2 \sin^2 \theta + 2ga \cos \theta + 2fb \sin \theta + c = 0 \qquad ...(2)$$

Suppose $$t = \tan \frac{\theta}{2};$$

then $$\cos \theta = \frac{1 - \tan^2 \theta/2}{1 + \tan^2 \theta/2} = \frac{1 - t^2}{1 + t^2}$$

and $$\sin \theta = \frac{2 \tan \frac{\theta}{2}}{1 + \tan^2 \theta/2} = \frac{1 - t^2}{1 + t^2}$$

Putting the values of cos θ and sin θ in (3), we get

$$a^2 \left(\frac{1 - t^2}{1 + t^2}\right)^2 + b^2 \left(\frac{2t}{1 + t^2}\right)^2 + 2ga \left(\frac{1 - t^2}{1 + t^2}\right) + 2fb \left(\frac{2t}{1 + t^2}\right) + c = 0$$

$$\Rightarrow \quad a^2 (1 - t^2)^2 + 4b^2t^2 + 2ga\,(1 - t^2)\,(1 + t^2) + 2fb\,2t \times (1 + t^2) + c\,(1 + t^2)^2 = 0$$

$$\Rightarrow \quad b^4 (a^2 - 2ga + c) + 4fbt^2 + t^2 (4b^2 - 2a^2 + 2c) + 4\,fbt + (a^2 + 2ga + c) = 0$$

If its roots be t_1, t_2, t_3 then we have

$$s_1 = \sum t_1 = \frac{-4fb}{a^2 - 2ga + c}, \; s_2 = \sum t_1 t_2 = \frac{4b^2 - 2a^2 + 2c}{a^2 - 2ga + c},$$

$$s_3 = \sum t_1\ t_2\ t_3 = \frac{-4fb}{a^2 - 2ga + c} \text{ and } s_4 = t_1\ t_2\ t_3\ t_4 = \frac{a^2 + 2ga + c}{a^2 - 2ga + c}.$$

Now $\tan \frac{1}{2}(\theta_1 + \theta_2 + \theta_3 + \theta_4) = \frac{s_1 - s_3}{a^2 - 2ga + c}$. [as $s_1 = s_2$]

Hence $\frac{\theta_1 + \theta_2 + \theta_3 + \theta_4}{2} = n\pi$

$\Rightarrow \quad \theta_1 + \theta_2 + \theta_3 + \theta_4 = 2n\pi =$ even multiplier π. **Proved.**

Example 46:

Q is the point on teh auxiliary circle corresponding to P on the ellipse; PLM is drawn parallel to CQ to meet the axes in L and M; prove that PL = b and PN = a.

Solution:

Let $\frac{x^2}{a^2} + \frac{y^2}{b^2} = 1$, be the equation of the ellipse and that of auxiliary circle be $x^2 + y^2 = a^2$.

Let the co-ordinates of P be $(a \cos \phi, b \sin \phi)$, ϕ is the eccentric angle of P.

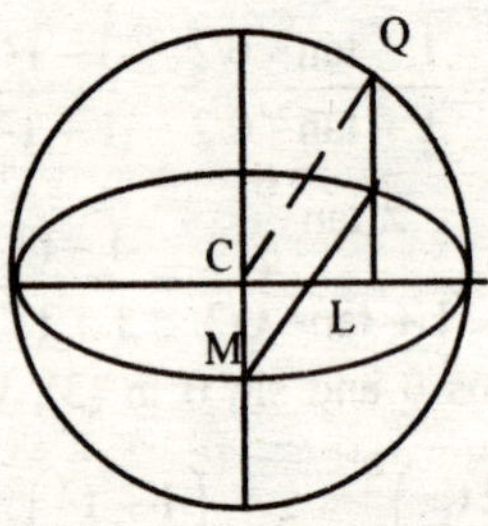

Draw PLM parallel to QC to meet the x-axis and y-axis in L and M respectively. Then angle PLX = ϕ. Now in parallelogram QPMC, PM = QC = a. If PN is the ordinate of P, i.e., = $b \sin \phi$, clearly

$PL = PN \operatorname{cosec} \phi = b \sin \phi \operatorname{cosec} \phi = b$. **Proved.**

Example 47:

Find the equation to the tangent and normal at the point (1, 4/3) of the ellipse $4x^2 + 9x^2 = 20$.

Solution:

The equation to the tangent at (1, 4/3) of the ellipse $4x^2 + 9y^2 = 20$ will be

$$4x.1 + 9y.4/3 = 20$$

or $$x + 3y = 5.$$

The slope of the tangent is –1/3 so that the slope of the normal will be 3.

So the equation of normal at point (1, 4/3) is

$$y - 4/3 = 3(x - 1)$$

or $$9x - 3y = 5.$$ **Ans.**

Example 48:

Find the equation to the tangent and normal at the point of the ellipse $5x^2 + 3y^2 = 137$ *whose ordinate is 2.*

Solution:

If the co-ordinates of the point be (h, 2), then as it lies on the ellipse $5x^2 + 3y^2 = 137$, the co-ordinates will satisfy it.

So $$5h^2 + 12 = 137$$

or $$5h^2 = 125; \qquad \therefore h = \pm 5.$$

Hence the co-ordinates of the points are (5, 2).

The tangent at (5, 2) will be $5.5x + 3.2y = 137$,

or $$25x + 6y = 137.$$

As its slope is –25/6, the slope of the normal will be 6/25.

So the equation of normal at (5, 2) is $(y - 2) = 6/25\ (x - 5)$

$$\Rightarrow \quad 6x - 25y + 20 = 0$$ **Ans.**

Similarly the equation of the tangent at

$(-5, 2)$ is $5x(-5) + 3y.2 = 137$

or $$-25x + 6y = 137$$

As its slope $= \frac{25}{6}$, so slope of the normal will be $\frac{6}{25}$.

Hence the equation to the normal at point (–5, 2) is

$$y - 2 = \frac{6}{25}(x + 5)$$

or $$6x + 25y - 20 = 0.$$ **Ans.**

Example 49(a):

If any ordinate MP meet the tangent at L in Q, prove that MQ and SP are equal.

Solution:

If $x^2/a^2 + y^2/b^2 = 1$, be the equation to the ellipse, the co-ordinates of L, the end point of the given latus rectum are $(-ae, b^2/a)$.

Tangent at L will be $\dfrac{-aex}{a^2} + \dfrac{b^2 y}{ab^2} = 1$

$$\Rightarrow \quad \frac{-ex}{a} + \frac{y}{a} = 1$$

$$\Rightarrow \quad y - ex = a. \qquad \text{...(1)}$$

Let the co-ordinates of any point P on the ellipse be (x_1, y_1); then equation to the ordinate PM is

$$x = x^2 \qquad \text{...(2)}$$

Solving (1) and (2), we get $y = a + ex_1$

Again $SP = a + ex_1$, where S is the focus, hence SP = MQ. **Proved.**

Example 49(b):

A tangent to the ellipse $\dfrac{x^2}{a^2} + \dfrac{y^2}{b^2} = 1$ *meets the ellipse* $\dfrac{x^2}{a} + \dfrac{y^2}{b} = a + b$ *in the points P and Q; prove that the tangents at P and Q are at right angles.*

Solution:

The equations to the ellipse are given as $\dfrac{x^2}{a^2} + \dfrac{y^2}{b^2} = 1$ ...(1)

and $\dfrac{x^2}{a^2} + \dfrac{y^2}{b^2} = a + b$ or $\dfrac{x^2}{a(a+b)} + \dfrac{y^2}{b(a+b)} = 1$ ...(2)

Let ϕ_1 and ϕ_2 be the eccentric angles of points P and Q; then their co-ordinates with respect to the ellipse (2) will be clearly

$$[\sqrt{\{a(a+b)\}} \cos\phi_1, \sqrt{\{b(a+b)\}} \sin\phi_1]$$

and $[\sqrt{\{a(a+b)\}} \cos\phi_2, \sqrt{\{b(a+b)\}} \sin\phi_2]$ respectively.

The equation to PQ will be

$$y - \sqrt{\{b(a+b)\}} \sin\phi_1 = \frac{\sqrt{b}(\sin\phi_1 - \sin\phi_2)}{\sqrt{a}(\cos\phi_1 - \cos\phi_2)} \times [x - \sqrt{\{a(a+b)\}} \cos\phi_1]$$

$$\Rightarrow y = -\sqrt{\frac{b}{a}} \cot \frac{\phi_1 + \phi_2}{2} x + \sqrt{[b(a+b)\}} \frac{\cos \frac{\phi_1 - \phi_2}{2}}{\sin \frac{\phi_1 + \phi_2}{2}} \quad ...(3)$$

If (3) touches the ellipse (1) also, it should be of the form $y = mx \sqrt{(a^2m^2 + b^2)}$.

Hence on comparing (2) and (4), we have $m = -\sqrt{\frac{b}{a}} \cot \frac{\phi_1 + \phi_2}{2}$ ...(5)

$$\text{and } a^2 m^2 + b^2 (a+b) \frac{\cos^2 \frac{\phi_1 - \phi_2}{2}}{\sin^2 \frac{\phi_1 + \phi_2}{2}} \quad ...(6)$$

Putting the value of m from (5) in (6), we get

$$\frac{a^2 b}{a} \cot^2 \frac{\phi_1 - \phi_2}{2} + b^2 = b(a+b) \frac{\cos^2 \frac{(\phi_1 - \phi_2)}{2}}{\sin^2 \frac{(\phi_1 + \phi_2)}{2}}$$

$$\Rightarrow a \cos^2 \frac{(\phi_1 + \phi_2)}{2} + b \sin^2 \frac{(\phi_1 - \phi_2)}{2} = (a+b) \cos^2 \frac{\phi_1 - \phi_2}{2}$$

$$\Rightarrow \frac{a}{2}[1 \cos(\phi_1 + \phi_2)] + \frac{b}{2} 1 - \cos(\phi_1 + \phi_2) = \frac{a+b}{2} \{1 + \cos(\phi_1 - \phi_2\}$$

$\because 2 \cos^2 \theta/2 = 1 + \cos \theta$ and $2 \sin^2 \theta/2 = (1 - \cos \theta)$

$\Rightarrow \quad (a+b) + (a-b) \cos(\phi_1 + \phi_2) = (a+b) + (a+b) \cos(\phi_1 - \phi_2)$

$\Rightarrow \quad (a-b) \cos(\phi_1 + \phi_2) = (a+b) \cos(\phi_1 - \phi_2)$

$\Rightarrow \quad (a-b)(\cos \phi_1 \cos \phi_2 - \sin \phi_1 \sin \phi_2)$

$= (a+b)(\cos \phi_1 \cos \phi_2 + \sin \phi_1 \sin \phi_2)$

$\Rightarrow \quad -2 \cos \phi_1 \cos \phi_2 = 2a \sin \phi_1 \sin \phi_2$ or $\cot \phi_1 \cot \phi_2 = -a/b$ ...(7)

The equation to tangent at P w.r.t. ellipse (2) is

$$\frac{x}{\sqrt{a(a+b)}} \cos \phi_1 + \frac{y}{\sqrt{[b(a+b)]}} \sin \phi_1 = 1$$

The slope of this tangent $= -\sqrt{(b/a)} \cot \phi_1$

Similarly slope of the tangent at Q is $= -\sqrt{(b/a)} \cot \phi_2$.

$\therefore$ Product of slopes $= (b/a) \cot \phi_1 \cot \phi_2 = \frac{b}{a} \times \left(-\frac{a}{b}\right) = -1$ from (7)

Hence the tangents at P and Q are at right angles. **Proved.**

Example 50:

Prove that the common tangent of the ellipses

$\frac{x^2}{a^2} + \frac{y^2}{b^2} = \frac{2x}{c}$ and $\frac{x^2}{b^2} + \frac{y^2}{a^2} + \frac{2x}{c} = 0$ *subtends a right angle at the origin.*

Solution:

The equation to the ellipse are given as

$$\frac{x^2}{a^2} + \frac{y^2}{b^2} = \frac{2x}{c} \text{ and } \frac{x^2}{b^2} + \frac{y^2}{a^2} + \frac{2x}{c} = 0$$

They may be written as

$$\left(\frac{x}{a} - \frac{a}{c}\right)^2 + \left(\frac{y^2}{b^2} - \frac{a^2}{c^2}\right) = 0 \text{ and } \left(\frac{x}{b} + \frac{b}{c}\right)^2 + \left(\frac{y^2}{a^2} - \frac{b^2}{c^2}\right) = 0.$$

Really the line y = ab/c, is a common tangent, since it cuts the two ellipse in coincident points whose co-ordinates are respectively $\left(\frac{a^2}{c}, \frac{ab}{c}\right)$ and $\left(\frac{-b^2}{b}, \frac{ab}{c}\right)$ say, P and Q and the co-ordinates of C are (0, 0).

Now slope of PC $= \frac{ab}{c} + \frac{a^2}{c} = \frac{b}{a} = m_1$ (say)

and slope of QC $= \frac{ab}{c} + \frac{-b^2}{c} = \frac{-a}{b} = m_2$ (say)

As $m_1 m_2 = \frac{b}{a}, \left(\frac{-a}{b}\right) = -1$ the common tangent substends right angle at the centre C.

Example 51:

The normal GP is produced to Q, so that GQ = n.GP.

Prove that the locus of Q is the ellipse $\frac{x^2}{a^2\,(n + e^2 - ne^2)^2} + \frac{y^2}{n^2 b^2} = 1.$

Solution:

Suppose P is any point on the ellipse

$$\frac{x^2}{a^2} + \frac{y^2}{b^2} = 1$$

whose co-ordinates are (a cos ϕ, b sin ϕ). The equation to normal PG at P is ax sec ϕ – by cosec ϕ = a^2 = b^2.

Solving it with y = 0, we get $x = \left(\frac{a^2 - b^2}{a}\right) \cos \phi$

$$\Rightarrow \quad x = \frac{a^2 e^2}{a} \cos \phi = ae^2 \cos \phi$$

Hence abscissa of G is $ae^3 \cos \phi$; so CG = $ae^2 \cos \phi$.

Draw PL and QM perpendiculars from P and Q respectively on x-axis. Then the triangles GPL and GQM are similar

$$\therefore \quad \frac{QM}{PL}\frac{GM}{GL} = \frac{GQ}{GP} = n, \text{ as } GQ = n.\ GP \text{ (given)}$$

$$\therefore \quad GM = n\ GL \text{ and } QM = n.PL$$

If the co-ordinates of Q be (x, y), then

$x = CM = CG + GM$ [where C is the centre (0, 0)]

$\Rightarrow \quad x = CG + n.GL = CG + n\,(CL - CG) = (1 - n)\,CG + CL.$

$= (1 - n)\, ae^2 \cos \phi + na \cos \phi = (e^2 - n^2 + n).\ a \cos \phi$ ...(1)

and $\quad y = QM = n.PN = nb \sin \phi$ ...(2)

By (1) and (2), we have $\cos \phi = \dfrac{x}{a(n + e^2 - ne^2)}$ and $\sin \theta = \dfrac{y}{ab}$.

Hence $\cos^2 \phi + \sin^2 \phi = \dfrac{x^2}{a^2\,(n + e^2 - ne^2)} + \dfrac{y^2}{n^2 b^2} = 1$

$$\Rightarrow \quad \frac{x^2}{a^2\,(n + e^2 - ne^2)} + \frac{y^2}{n^2 b^2} = 1$$ **Proved.**

Example 52:

Show that the angle between the tangents to the ellipse $\frac{x^2}{a^2} + \frac{y^2}{b^2} = 1$ *and the circle* $x^2 + y^2 = ab$ *at their points of intersection is*

$$\tan^{-1} \frac{a - b}{\sqrt{ab}}$$

Solution:

As the equation to the ellipse is $\frac{x^2}{a^2} + \frac{y^2}{b^2} = 1$ ...(1)

and the equation to the circle is $x^2 + y^2 = ab$.

To find the co-ordinates of points or intersection, we have to solve (1) and (2). From (2) $y^2 = ab - x^2$.

Putting in (1), $\dfrac{x^2}{a^2} + \dfrac{ab - x^2}{b^2} = 1$

$$\Rightarrow \quad b^2x^2 + a^2(ab - x^2) = a^2b^2$$

$$\Rightarrow \quad x^2(b^2 + a^2) = -a^3b + a^2b^2 = a^2b(b - a);$$

$$\therefore \quad x = \pm\sqrt{\left(\frac{a^2b}{a+b}\right)} \text{ and hence } y = \pm\sqrt{\left(\frac{ab^2}{b+b}\right)} \text{ from (3)}$$

$\therefore$ The co-ordinates of one of the points of intersection (say P) may be

taken as $\left\{\sqrt{\left(\dfrac{a^2b}{a+b}\right)}, \sqrt{\left(\dfrac{ab^2}{b+b}\right)}\right\}$

Equation of tangent at P w.r.t. (1) is given as

$$\frac{x}{a^2}\sqrt{\left(\frac{a^2b}{a+b}\right)} + \frac{y}{b^2}.\sqrt{\left(\frac{ab^2}{b+b}\right)} = 1$$

$$\text{Its slope} = -\frac{\text{coefficient of x}}{\text{coefficient of y}} = -\sqrt{\left(\frac{a}{a+b}\right)}\frac{1}{a^2} \times b^2\sqrt{\left(\frac{a+b}{b}\right)}$$

$$= -\frac{b^{\frac{3}{2}}}{a^{\frac{3}{2}}} = m_1 \text{ (say).}$$

The equation to tangent at P. w.r.t. (2) is

$$x\sqrt{\left(\frac{a^2b}{a+b}\right)} + y\sqrt{\left(\frac{ab^2}{a+b}\right)} = ab.$$

$$\text{It slope} = -\sqrt{\left(\frac{a^2b}{a+b}\right)} \times \sqrt{\left(\frac{a+b}{ab^2}\right)} = \frac{a^{\frac{1}{2}}}{b^{\frac{1}{2}}} = m_2 \text{ (say).}$$

Hence if θ be the angle between these two tangent at P.

Then we have

$$\tan\theta \frac{m_1 - m_2}{1 + m_1m_2} = \frac{-\dfrac{b^{3/2}}{a^{3/2}} + \dfrac{a^{1/2}}{b^{1/2}}}{1 + \dfrac{b^{3/2}}{a^{3/2}} + \dfrac{a^{1/2}}{b^{1/2}}} = \frac{a^2 - b^2}{a^{3/2}b^{1/2} + b^{3/2}a^{1/2}}$$

$$\theta = \tan^{-1} \frac{(a-b)(a+b)}{a^{1/2}\, b^{1/2}\,(a+b)} = \tan^{-1} \frac{a-b}{\sqrt{(ab)}}$$ **Hence proved.**

Example 53(a):

If the straight line $y = mx + c$ meet the ellipse proved the equation to the circle, described on the line joining the points of intersection as diameter is

$$(a^2m^2 + b^2)(x^2 + y^2) + 2ma^2cx - 2b^2cy + c^2(a^2 + b^2) - a^2b^2(1 + m^2) = 0$$

Solution:

The line is given as $y = mx + c$ ...(1)

and the ellipse is given as $x^2/x^2 + y^2/b^2 = 1.$ ...(2)

Solving (1) and (2), we get $\frac{x^2}{a^2} + \frac{(mx+c)^2}{b^2} = 1$

$\Rightarrow\ b^2x^2 + a^2(m^2x^2 - 2mcx + c^2) = a^2b^2$

$\Rightarrow\ (a^2m^2 + b^2)x^2 + 2a^2mcx + a^2c^2 - a^2b^2 = 0$

Let x_1 and x_2 be the two roots of this equation

$$\therefore \quad x_1 + x_2 = \frac{-2a^2\,mc}{a^2m^2 + b^2} \quad ...(3)$$

$$\text{and} \quad x_1 x_2 = \frac{a^2c^2 - a^2b^2}{a^2m^2 + b^2} \quad ...(4)$$

Let y_1 and y_2 be the corresponding ordinates for the abscissa x_1 and x_2; so the co-ordinates of the points of intersection will be (x_1, y_1) and (x_2, y_2). As these lie on line $y = mx + c$, we have $y_1 = mx_1 + c$ and $y_2 = mx_2 + c$

whence $y_1 + y_2 = m(x_1 + x_2) + 2c$ ...(5)

and $y_1y_2 = (mx_1 + c)(mx_2 + c) = m^2x_1x_2 + cm(x_1 + x_2) + c^2$. ...(6)

The equation to the circle drawn with line joining (x_1, y_1) and (x_2, y_2) as diameter is given as

$$(x - x_1)(x - x_2) + (y - y_1)(y - y_2) = 0$$

$$\Rightarrow\quad x^2 + y^2 - x(x_1 + x_2) - y(y_1 + y_2) + x_1x_2 + y_1y_2 = 0$$

$$\Rightarrow\quad x^2 + y^2 - x(x_1 + x_2) - y\{m(x_1 + x_2) + 2c\} + x_1x_2 + m^2x_1x_2 + cm(x_1 + x_2) + c^2 = 0 \text{ [by (5) and (6)]}$$

Putting the values from (3) and (4), we get

$$x^2 + y^2 - x\left(\frac{-2a^2 mc}{a^2 m^2 + b^2}\right) - y\left(\frac{-2a^2 m^2 c}{a^2 m^2 + b^2} + 2c\right) + \frac{a^2 c^2 - a^2 b^2}{a^2 m^2 + b^2}$$

$$+ \frac{m^2 (a^2 c^2 - a^2 b^2)}{a^2 m^2 + b^2} + cm \frac{(2a^2 mc)}{a^2 m^2 + b^2} + c^2 = 0$$

$$\Rightarrow \quad (a^2m^2 + b^2)(x^2 + y^2) + 2a^2 mcx - 2b^2cy + c^2 (a^2 + b^2)$$

$$- a^2b^2 (1 + m^2) = 0.$$ **Proved.**

Example 53(b):

Two tangents to the ellipse intersect at right angles; prove that the sum of the squares of the chords which the auxiliary circle intercepts on them is constant, and equal to the square on the line joining the foci.

Solution:

$x^2/a^2 + y^2/b^2 = 1$, be the equation to the ellipse then $x^2 + y^2 = a^2$ will be that of the auxiliary circle.

If PQ be any tangent on the ellipse, then its equation may be taken as $y = mx + \sqrt{(a^2m^2 + b^2)}$.

Length of the perpendicular CM on the chord PQ $= \dfrac{\sqrt{(a^2 m^2 + b^2)}}{\sqrt{(1 + m^2)}}$

As M will be the mid-point of the chord PQ.

So $\qquad$ QM = MP

$\Rightarrow PQ^2 = (2MP)^2 = 4MP^2 = 4 (CP^2 - CM^2)$ (as CM perpendicular PQ)

$$\Rightarrow PQ^2 = 4\left\{a^2 - \left(\frac{a^2 m^2 + b^2}{1 + m^2}\right)\right\} = \frac{4\{a^2 - b^2\}}{1 + m^2} = \frac{4a^2 c^2}{1 + m^2} \qquad ...(1)$$

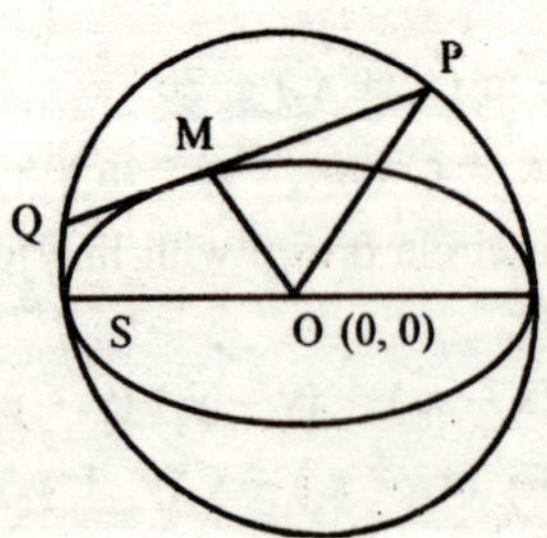

The square of the perpendicular on the other chord which is perpendicular to PQ can be obtained by putting $-1/m$ for m in (1)

Hence the square of another perpendicular

$$\frac{4a^2 e^2}{1 + 1/m^2} = \frac{4a^2 e^2 m^2}{1 + m^2} \qquad ...(2)$$

Adding (1) and (2), we have

$$= \frac{4a^2 e^2}{1 + m^2} + \frac{4a^2 e^2 m^2}{1 + m^2} = \frac{4a^2 e^2 (1 + m^2)}{(1 + m^2)} = 4a^2 e^2 = (2ae)^2 = (SS')^2$$

Example 54:

Prove that PG, Pg = SP. SP. and CG.CT = CS.

Solution:

Take the equation to the ellipse as $\frac{x^2}{a^2} + \frac{y^2}{b^2} = 1$, so the co-ordinate of any point P on it will be (a cos ϕ, b sin ϕ). ...(1)

Solving with y = 0, we get x $\frac{(a^2 - b^2)}{a} \cos\phi$

Hence the co-ordinates of G are $\left\{\frac{(a^2 - b^2)}{a} \cos\phi, 0\right\}$

Solving (1) with y-axis, i.e., x = 0, y = $\frac{b^2 - a^2}{b} \sin\phi$

So the co-ordinates of g are $\left(0, \frac{b^2 - a^2}{a} \sin\phi\right)$

$$\therefore \; PG^2 = \left(a\cos\phi \; \frac{a^2 - b^2}{a} \cos\phi\right) + b^2 \sin^2\phi = \left(\frac{b^2}{a}\cos\phi\right)^2 + b^2 \sin^2\phi$$

$$\Rightarrow \qquad PG = \frac{b}{a}\sqrt{(b^2 \cos^2\phi + a^2 \sin^2\phi)}$$

and $\qquad Pg^2 = a^2 \cos^2\phi + \left(b\sin\phi - \frac{b^2 - a^2}{b} \sin\phi\right)^2$

$$\Rightarrow \; Pg = \sqrt{\left(a^2 \cos^2\phi + \frac{a^2}{b^2} \sin^2\phi\right)} = \frac{a}{b}\sqrt{(b^2 \cos^2\phi + a^2 \sin^2\phi)}$$

$$\therefore \qquad PG.Pg = b^2 \cos^2\phi + a^2 \sin^2\phi \qquad ...(2)$$

Again PS = (a + ea cos ϕ) and PS' = (a – ea cos ϕ).

So PS.PS' = (a + ea cos ϕ) (a – ea cos ϕ) = a2 – $a^2 e^2 \cos^2\phi$.

Putting the value of e^2 from $e^2 = (a^2 - b^2)/a^2$, we get

$$PS.PS' = a^2 - a^2 \frac{(a^2 - b^2)}{a^2} \cos^2 \phi = a^2 - (a^2 - b^2) \cos^2 \phi$$

$$= a^2 - a^2 \cos^2 \phi + b^2 \cos^2 \phi = a^2 \sin^2 \phi + b^2 \cos^2 \phi. \quad ...(3)$$

From (2) and (3), PG, Pg = PS.PS'.

Again equation to tangent PT (where T is the point of intersection with x-axis) is

$$\frac{x}{a} \cos \phi + \frac{y}{b} \sin \phi = 1$$

Solving with x-axis, i.e., y = 0, we get, $x = a \sec \phi$.

Hence the co-ordinates of T are $(a \sec \phi, 0)$.

$$\therefore \quad CG.CT = \frac{a^2 - b^2}{a} \cos \phi . \ a \sec \phi = a^2 = b^2 = a^2 e^2 = CS^2$$ **Proved.**

Example 55(a):

Prove that the circle on any focal distance as diameter touches the auxilliary circle.

Solution:

Let any point P be $(a \cos \phi, b \sin \phi)$ on the ellipse

$$\frac{x^2}{a^2} + \frac{y^2}{b^2} = 1.$$

So the focus S will be (–ae, 0). The equation to the circle drawn with PS as diameter is

$$(x - a \cos \phi)(x + ae) + (y - b \sin \phi)(y) = 0$$

$$\Rightarrow \quad x^2 + y^2 + x(ae - a \cos \phi) - by \sin f = a^2 e \cos \phi = 0 \quad ...(1)$$

and equation tot eh auxiliary circle is $x^2 + y^2 = a^2$. ...(2)

The equation of radical axis of (1) and (2) will be

$$ax(e - \cos \phi) - by \sin \phi + a^2 - a^2 e \cos \phi) = 0 \quad ...(3)$$

The radical axis must touch the circle (2), if length of perpendicular from (0, 0) on (3) = radius of that circle.

Length of perpendicular from (0, 0) on (3)

$$= \frac{a^2 - a^2 e \cos \phi}{\sqrt{[a^2 (e - \cos \phi)^2 + b^2 \sin^2 \phi]}}$$

$$= \frac{a^2(1 - e\cos\phi)}{\sqrt{[a^2(e^2 + \cos^2\phi - 2e\cos\phi) + a^2(1 - e^2)\sin^2\phi]}}$$

$$= \frac{a^2(1 - e\cos\phi)}{\sqrt{e^2[1 - \sin^2\phi) + (\cos^2\phi + \sin^2\phi) - 2e\cos\phi]}}$$

$$= \frac{a(1 - e\cos\phi)}{\sqrt{[e^2\cos^2\phi + 1 - 2e\cos\phi]}} = \frac{a(1 - e\cos\phi)}{\sqrt{[(1 - e\cos\phi)^2]}}$$

$$= \frac{a(1 - e\cos\phi)}{1 - e\cos\phi} = a = \text{radius of the circle.}$$

Example 55(b):

If a number of ellipses be described having the same major axis, but a variable minor axis, prove that the tangents at the ends of their latus rectum pass through one or other of two fixed points.

Solution:

Suppose the equation to any of the ellipse be

$$\frac{x^2}{a^2} + \frac{y^2}{b^2} = 1,$$

then the ends of the latus rectum are ($\pm ae, \pm b^2/a$). The tangent at this point will be

$$\frac{\pm aex}{a^2} + \frac{\pm b^2 y}{ab^2} = 1 \quad \text{or } \pm ex \pm y = a \qquad ...(1)$$

(1) represents four tangents corresponding to four points and it always passes through one of the points ($0, \pm a$), whatever the value of e may be.

Example 55(c):

The tangent at P meets the axes in T and t, and CY is the perpendicular on it from the centre; prove that (1) Tt. PY = $a^2 - b^2$, and (2) the least value of Tt is a + b.

Solution:

Suppose P ≡ ($a\cos\phi$, $b\sin\phi$) be the point on the ellipse

$$\frac{x^2}{a^2} + \frac{y^2}{b^2} = 1$$

the tangent PT meets the x-axis at T and y-axis at t. The equation of tangent PT is given as

$$\frac{x}{a}\cos\phi + \frac{y}{b}\sin\phi = 1 \qquad ...(1)$$

As in last question, the co-ordinates of T ≡ (a sec φ, 0).

Solving (1) with y-axis, i.e., x = 0, we get co-ordinates of t as (0, b cosec φ).

Equation of CY perpendicular to (1) will be

$$ax\sin\phi - by\cos\phi = 0 \qquad ...(2)$$

PY will be the length of perpendicular from P on (2); so

$$PY = \frac{a^2\sin\phi\cos\phi - b^2\sin\phi\cos\phi}{\sqrt{(a^2\sin^2\phi + b^2\cos^2\phi)}}$$

$$\Rightarrow \qquad PY = \frac{(a^2 - b^2)\sin\phi\cos\phi}{\sqrt{(a^2\sin^2\phi + b^2\cos^2\phi)}}$$

and $Tt = \sqrt{\{(a\sec\phi)^2 + (b\,\text{cosec}\,\phi)^2\}} = \dfrac{\sqrt{(a^2\sin^2\phi + b^2\cos^2\phi)}}{\sin\phi\cos\phi}$

(a) $Tt\,PY = \dfrac{\sqrt{(a^2\sin^2\phi + b^2\cos^2\phi)}}{\sin\phi\cos\phi} \cdot \dfrac{(a^2\,b^2)\sin\phi\cos\phi}{\sqrt{(a^2\sin^2\phi + b^2\cos^2\phi)}} = a^2 - b^2$

(b) $Tt^2 = a^2\sec^2\phi + b^2\,\text{cosec}^2\phi = a^2(1 + \tan^2\phi) + b^2(1 + \cot^2\phi)$

$$= (a^2 + b^2) + (a^2\tan^2\phi + b^2\cot^2\phi)$$

$$= (a^2 + b^2 + 2ab) + (a^2\tan^2\phi + b^2\cot^2\phi - 2ab)$$

$$= (a + b)^2 + (a\tan\phi - b\cot\phi)^2.$$

So the least value of Tt is when (a tan φ – b cot φ) = 0 and this value is (a + b).

Example 55(d):

Prove that the locus of the middle points of the portions of tangents included between the axes is the curve $a^2/x^2 + b^2/y^2 = 4$.

Solution:

We know that $y = mx + \sqrt{(a^2m^2 + b^2)}$ is always tangent on the ellipse $x^2/a^2 + y^2/b^2 = 1$. This tangent will meet the x-axis, i.e., y = 0, at the point whose co-ordinates are $\left(\dfrac{-\sqrt{(a^2\ m^2 + b^2)}}{m}, 0\right)$ and also it will meet the y-axis, i.e., x = 0 at a point where $y = \sqrt{(a^2 m^2 + b^2)}$.

If (x, y) be the co-ordinates of the middle points of the line joining these two points then

$$x = \frac{-\sqrt{(a^2\ m^2 + b^2)}}{2m} \qquad ...(1)$$

$$y = \frac{-\sqrt{(a^2\ m^2 + b^2)}}{2} \qquad ...(2)$$

Dividing (1) by (2), we get m = –y/x. Substituting for 'm' in (2)

$$y = \frac{1}{2}\sqrt{\left(\frac{a^2y^2}{x^2} + b^2\right)} \text{ or } 4y = \frac{a^2y^2}{x^2} + b^2 \text{ (on squaring)}$$

or $a^2/x^2 + b^2/y^2 = 4$. **Proved.**

Example 56:

Prove that the locus of the feet of the perpendicular drawn from the centre upon any tangent to the ellipse is $r^2 = a^2 \cos^2 \theta + b^2 \sin^2 \theta$.

Solution:

Suppose the equation to tangent to the ellipse be

$$x \cos \alpha + y \sin \alpha = p \qquad ...(1)$$

Comparing (1) with the general equation of the tangent,

$$y = mn + \sqrt{(a^2m^2 + b^2)}, \text{ we get } m = -\cot \alpha \qquad ...(2)$$

and also $a^2m^2 + b^2 = p^2 \operatorname{cosec}^2 \alpha$. ...(3)

Eliminating m between (2) and (3), putting the value of m from (2) in (3), we get $a^2 \cos^2 a + b^2 \sin^2 a = p^2$.

As p is the length of the perpendicular on the line

$$x \cos \alpha + y \sin \alpha = p$$

from the origin (0, 0) therefore the polar co-ordinates of foot of the perpendicular are (p, α).

Hence generalising for p and α, the required locus is

$$a^2 \cos^2 \theta + b^2 \sin^2 \theta = r^2.$$ **Proved.**

Example 57(a):

Find the tangent of the angle between OP and the normal at P, and prove that its greatest value is $(a^2 - b^2)/2ab$.

Solution:

If θ be the angle between the normal PG at $P \equiv (a \cos \phi, b \sin \phi)$ and PC where C is the centre of the ellipse given by $\frac{x^2}{a^2} + \frac{y^2}{b^2} = 1$, then the equation to the normal PG is $ax \sec \phi - by \operatorname{cosec} \phi = a^2 - b^2$.

$$\text{It slope} = \frac{-a \sec \phi}{-b \operatorname{cosec} \phi} = \frac{a}{b} \tan \phi = m_1 \text{ (say)}.$$

$$\text{The slope of PC} = \frac{b \sin \phi}{a \cos \phi} = \frac{b}{a} \tan \phi = m_2 \text{ (say)}.$$

$$\tan \theta = \frac{m_1 - m_2}{1 + m_1 m_2} = \frac{(a/b) \tan \phi - (b/a) \tan \phi}{1 + (a/b) \tan \phi \times (b/a) \tan \phi} = \frac{(a^2 - b^2)}{ab(1 + \tan^2 \phi)}$$

$$= \frac{(a^2 - b^2)}{2ab} \cdot \frac{2 \tan \phi}{1 + \tan^2 \phi} = \frac{a^2 - b^2}{2ab} \sin 2\phi \left\{ \text{as } \sin 2\phi = \frac{2 \tan \phi}{1 + \tan^2 \phi} \right\}$$

The value of $\tan \theta$ will be maximum when $\sin 2\phi$ is maximum i.e., $\sin 2\phi$ is unity. Therefore the greatest value of $\tan \theta$ is $(a^2 - b^2)/2ab$. **Proved.**

Example 57(b):

The normal at P meets the axes in G and g; show that the loci of the middle points of PG and Gg are respectively the ellipses

$$\frac{4x^2}{a^2 (1 + e^2)^2} + \frac{4y^2}{b^2} = 1, \text{ and } a^2x^2 + b^2y^2 = \frac{1}{4} (r^2 - b^2)$$

Solution:

The equation to normal at any point $P \equiv (a \cos \phi, b \sin \phi)$ on the ellipse $x^2/a^2 + y^2/b^2 = 1$ is

$$ax \sec \phi - by \operatorname{cosec} \phi = a^2 - b^2. \qquad \text{...(1)}$$

Let it meet the axis at G and g respectively.

Solving (1) with x-axis, i.e., $y = 0$, we get $x = \frac{a^2 - b^2}{a} \cos \phi$

So the co-ordinates of G are $= \left(\frac{a^2 - b^2}{a} \cos \phi, 0 \right)$

Solving (1) with y-axis, i.e., x = 0, we get the co-ordinates of g as $\left(0, \frac{a^2 - b^2}{a} \sin\phi\right)$.

(a) If the co-ordinates of the middle point of PG be (x, y), then

$$x = \frac{1}{2}\left[a\cos\phi + \frac{a^2 - b^2}{a}\cos\phi\right] \text{ and } y = \frac{b\sin\phi}{2}$$

$$\text{Hence } \cos\phi = \frac{2xa}{2a^2 - b^2} = \frac{2xa}{2a^2 - a^2(1 - c^2)} = \frac{2xa}{a^2 + a^2e^2} = \frac{2x}{a(1 + e^2)}$$

and $$\sin\phi = \frac{2y}{b}$$

Now squaring and adding, we get $\frac{4x^2}{a^2(1 - e^2)^2} + \frac{4y^2}{b^2} = \cos^2\phi + \sin^2\phi$

$\Rightarrow \quad \frac{x^2}{a^2(1 + e^2)^2} + \frac{4y^2}{b^2} = 1$ **Proved.**

(b) If the co-ordinates of the middle point of Gg be (x, y), then

$$x = \frac{a^2 - b^2}{2a}\cos\phi \text{ and } y = \frac{a^2 - b^2}{-2b}\sin\phi.$$

$$\text{Hence } \cos\phi = \frac{2ax}{a^2 - b^2} \text{ and } \sin\phi = \frac{-2by}{a^2 - b^2}$$

On squaring and adding, we get

$$\frac{4a^2x^2}{(a^2 - b^2)^2} + \frac{4b^2y^2}{(a^2 - b^2)} = \cos^2\phi + \sin^2\phi = 1$$

$\Rightarrow \quad a^2x^2 + b^2y^2 = \frac{1}{4}(a^2 - b^2)$ **Proved.**

Example 57(c):

Tangents are drawn, from the point (3, 2) to the ellipse $x^2 + 4y^2 = 9$. Find the equation to their chord of contact and the equation of the straight line joining (3, 2) to the middle point of this chord of contact.

Solution:

The equation of the chord of contact of the point (3, 2) with respect to the ellipse $x^2 + 4y^2 = 9$ is

$3.x + 4y.2 = 9$ or $3x + 8y = 9$. **Ans.**

Slope of the chord of contact $= -\frac{3}{8} = m$ (say).

The required line joining the middle points of the chord to (3, 2) is the diameter of the chord of contact.

As equation to the diameter is $y = \frac{-b^2}{a^2m}x$ ...(1)

(here $a^2 = 9$, $b^2 = 9/4$ and $m = -3/8$)

Hence the required line is $y = \frac{-9}{4 \times 9} \times \frac{-8}{3}x = \frac{2}{3}x$ or $3y = 2x$. **Ans.**

Example 57(d):

Prove that a chord which joins the ends of a pair of conjugate diameters of an ellipse always touches a similar ellipse.

Solution:

Suppose CP and CQ are the conjugate diameters of the ellipse $x^2/a^2 + y^2/b^2 = 1$, and C be the centre. Now, if P be any point $(a \cos \theta, b \sin \theta)$ then the co-ordinates of Q will be $\{a \cos (90° + \theta), be \sin (90° + \theta)\}$ or $(-a \sin \theta, b \cos \theta)$.

$$y - b \cos \theta = \frac{b (\sin \theta - \cos \theta)}{a (\cos \theta + \sin \theta)} (x + a \sin \theta)$$

$$\Rightarrow \quad ay (\cos \theta + \sin \theta) - bx (\sin \theta - \cos \theta)$$

$$= ab \{\sin \theta \cos \theta + \sin^2 \theta - \sin \theta \cos \theta + \cos^2 \theta\} = ab$$

$$\Rightarrow \quad (y/b) (\cos \theta + \sin \theta) + (x/a) (\cos \theta - \sin \theta) = 1$$

$$\Rightarrow \quad (x/a) \cos (\theta + \pi/4) + (y/b) \sin (\theta + \pi/4) = 1/\sqrt{2} \quad ...(1)$$

(1) is always tangent to an ellipse $x^2/a^2 + y^2/b^2 = 1/2$ and which is similar to the given ellipse $x^2/a^2 + y^2/b^2 = 1$. **Proved.**

Example 58(a):

Any ordinate NP of an ellipse meets the auxiliary circle in Q; prove that the locus of the intersection of the normals at P and Q is the circle

$$x^2 + y^2 = (a + b)^2.$$

Solution:

Suppose the equation to the ellipse be $x^2/a^2 = y^2/a^2 = 1$; therefore the equation to the auxiliary circle will be $x^2 + y^2 = a^2$.

Let the co-ordinates of any point P be $(a \cos \phi, a \sin \phi)$; then those of Q will be $(a \cos \phi, a \sin \phi)$.

Normal at P is $ax \sec \phi - by \operatorname{cosec} \phi = a^2 - b^2$ ...(1)

and normal at Q will be $y = x \tan \phi$. ...(2)

Now, we have to eliminate ϕ between (1) and (2).

$$\text{Hence } \sec \phi = \sqrt{(1 + \tan^2 \phi)} = \sqrt{\left(1 + \frac{y^2}{x^2}\right)} = \sqrt{\left(\frac{x^2 + y^2}{x^2}\right)}$$

$$\text{and } \operatorname{cosec} \phi = \frac{\sec \phi}{\tan \phi} = \sqrt{\left(\frac{x^2 + y^2}{x}\right)} \Big/ \frac{y}{x} = \sqrt{\frac{(x^2 + y^2)}{y}}$$

Putting the values in (1), we get

$$ax \sqrt{\frac{(x^2 + y^2)}{x}} - by \sqrt{\frac{(x^2 + y^2)}{y}} = a^2 - b^2$$

$\Rightarrow$ $(a - b). \sqrt{(x^2 + y^2)} = a^2 - b^2$

or $\sqrt{(x^2 + y^2)} = a + b$

$\Rightarrow$ $x^2 + y^2 = (a + b)^2$. **Proved.**

Example 58(b):

In the ellipse $\frac{x^2}{36} + \frac{y^2}{9} = 1$, *find the equation to the chord which passes through the point (2, 1) and is bisected at that point.*

Solution:

The equation of the polar of (2, 1) w.r.t. the given ellipse

$$\frac{x^2}{36} + \frac{y^2}{9} = 1 \text{ is } \frac{2x}{36} + \frac{y}{9} = 1$$

$\therefore$ Slope of the polar is (–1/2).

The slope of the required chord whose middle point is (2, 1) will also be (–1/2). Hence its equation is

$(y - 1) = -1/2\ (x - 2)$

or $x + 2y = 4 = 0$. **Ans.**

Example 58(c):

Find with respect to the ellipse $4x^2 + 7y^2 = 8$.

(1) the polar of the point $(-1/2, 1)$, *and*

(2) the pole of the straight line $12x + 7y + 16 = 0$.

Solution:

The equation of the given ellipse is $4x^2 + 7y^2 = 8$. ...(1)

(a) The equation to the polar of the point (–1/2, 1) w.r.t. (1) will be $4x(-1/2) + 7y(1) = 8$ or $2x - 7y + 8 = 0$.

(b) If the pole of the given line $12x + 7y + 16 = 0$ ...(2)

be (x_1, y_1), then the equation to the polar of (x_1, y_1) w.r.t. (1) will be

$$4xx_1 + 7yv_1 - 8 = 0 \quad ...(3)$$

As the lines (2) and (3) represent the same line, comparing the coefficients, we have

$$\frac{4x_1}{12} = \frac{7y_1}{7} = \frac{-8}{16}$$

Solving, we get $x_1 = -\frac{3}{2}$ and $y_1 \frac{1}{2}$

So the pole is $\left(-\frac{3}{2}, -\frac{1}{2}\right)$ **Ans.**

Example 59:

If the pole of the normal at P lie on the normal at Q, then show that the pole of the normal at Q lies on the normal at P.

Solution:

If the co-ordinates of P be (a cos α, b sin α); then equation of the normal on P with respect to the ellipse

$$\frac{x^2}{a^2} + \frac{y^2}{b^2} = 1 \quad ...(1)$$

is $\quad$ $ax \sec a - by \operatorname{cosec} a = a^2 - b^2$ ...(2)

If the pole of (2) be (x_1, y_1) (say L), then the equation to the polar of (x_1, y_1), w.r.t. (1) is given as

$$\frac{xx_1}{a^2} + \frac{yy_1}{b^2} = 1 \quad ...(3)$$

(2) and (3) will be similar if

$$\frac{x_1}{a^2.a \sec \alpha} = \frac{y_1}{-b^2.b \operatorname{cosec} \alpha} = \frac{1}{a^2 - b^2}$$

So $\quad x_1 = \frac{a^3 \sec \alpha}{a^2 - b^2}$ and $y_1 \frac{b^3 \operatorname{cosec} \alpha}{a^2 - b^2}$

$\therefore$ The co-ordinates of the pole L are $\left(\dfrac{a^3 \sec\alpha}{a^2 - b^2}, \dfrac{-b \operatorname{cosec}\alpha}{a^2 - b^2}\right)$

Again the equation to the normal at a point Q whose co-ordinates are $(a\cos\beta, b\sin\beta)$ is $ax\sec\beta - by\operatorname{cosec}\beta = a^2 - b^2$. ...(4)

By hypothesis, the point L, whose co-ordinates are

$\left(\dfrac{a^3 \sec\alpha}{a^2 - b^2}, \dfrac{-b \operatorname{cosec}\alpha}{a^2 - b^2}\right)$ lies on (4).

So $$\frac{a^4 \sec\alpha \sec\beta}{a^2 - b^2} + \frac{b^4 \operatorname{cosec}\alpha \operatorname{cosec}\beta}{a^2 - b^2} = a^2 - b^2$$

This condition is symmetrical in α and β. Similarly, the same condition is again obtained when the pole of the normal at Q is made to lie on the normal at P. Hence the result. **Proved.**

Example 60:

If a fixed straight line parallel to either axis meet a pair of conjugate diameters in the points K and L, show that the circle described on KL, as diameter passes through two fixed points on the other axis.

Solution:

Suppose the conjugate diameters are given by

$$b^2x^2 - pxy - a^2y^2 = 0, \quad ...(1)$$

and the line be $\quad y = c \quad ...(2)$

As (1) meets the line $y = c$ at points K and L, its abscissae will be given by the equation $b^2x^2 - pex - a^2c^2 = 0$.

Let the roots be x_1 and x_2 : $x_1 + x_2 = \dfrac{pc}{b^2}$ and $x_1 x_2 = \dfrac{a^4 c^2}{b^2}$

So the co-ordinates K and L will be (x_1, c) and (x_2, c). Now equation to the circle drawn with KL as diameter as

$$(x - x_1)(x - x_2) = (y - c)(y - c) = 0$$

$$\Rightarrow \quad x^2 + y^2 = x(x_1 + x_2) - 2cy + x_1x_2 + c^2 = 0$$

$$\Rightarrow \quad x^2 + y^2 - \frac{pcx}{2} - 2cy\,\frac{a^4 c^2}{b^2} + c^2 = 0 \quad ...(3)$$

(on substituting the values of $x_1 + x_2$ and x_1x_2)

Equation No. (3) meets $x = 0$ in points given by

$$y^2 - 2cy + c^2 = a^2c^2/b^2.$$

$\Rightarrow$ $$(y - c)^2 = (ac/b)^2$$

or $$y = c \pm ac/b$$

Therefore, the co-ordinates of the point are (0, c + ac/b) and (0, c – ac/b). These points are fixed as a, b and c are constants. **Proved.**

Example 61(a):

Write down the equation of the pair of tangents drown to the ellipse $3x^2 = 2y^2 = 5$ from the point (1, 2), and prove that the angle between them is $\tan^{1} 12\sqrt{5}/5$.

Solution:

The equation to the given ellipse is 3x2 + 2y2 – 5 = 0 and the co-ordinates of the point are (1, 2). Here

$$S \equiv 3x^2 + 2y^2 - 5 \text{ and } S_1 \equiv 3\,(1)^2 + 2\,(2)^2 - 5 = 6$$

and $T \equiv 3x.1 + 2y.2 - 5$ [i.e., tangent at (1, 2) as if (1, 2) is on the ellipse] or $T \equiv 3x + 4y - 5$.

The equation to the pair of tangents is $SS_1 = T^2$.

So the equation is $(3x^2 + 2y^2 - 5)\,5 = (3x + 4y - 5)^2$.

$\Rightarrow$ $$9x^2 - 24xy - 4y^2 + 30x + 40y - 55 = 0.$$

The angle between the lines is given by $\tan\theta = \dfrac{2\sqrt{(180)}}{5} = \dfrac{12\sqrt{5}}{5}$.

$\therefore$ $$\theta = \tan^{-1}\left(\frac{12}{5} = \sqrt{5}\right)$$ **Ans.**

Example 61(b):

Show that the four lines which join the foci to two points P and Q on an ellipse all touch a circle whose centre is the pole of PQ.

Solution:

Taking the co-ordinates of P and Q of the ellipse as $(a\cos\phi_1, b\sin\phi_1)$ and $(a\cos\phi_2, b\sin\phi_2)$ respectively, the equation to line PQ is

$$\frac{x}{a}\cos\frac{\phi_1+\phi_2}{2} + \frac{y}{b}\sin\frac{\phi_1+\phi_2}{2} = \cos\frac{\phi_1-\phi_2}{2} \qquad ...(1)$$

If the pole of (1) be (x_1, x_2); then equation to the polar of (x_1, y_1) with respect to the ellipse $\dfrac{x^2}{a^2} + \dfrac{y^2}{b^2} = 1$ is given by

$$\frac{xx_1}{a^2} + \frac{yy_1}{b^2} = 1 \qquad ...(2)$$

Since the two lines (1) and (2) represent the same line, so the coefficients should be proportional, hence, we have

$$\frac{x_1}{a \cos \frac{(\phi_1 + \phi_2)}{2}} = \frac{y_1}{b \sin \frac{(\phi_1 + \phi_2)}{2}} = \frac{1}{\cos \frac{(\phi_1 + \phi_2)}{2}}$$

$$\Rightarrow \quad x_1 = \frac{a \cos\left(\frac{\phi_1 + \phi_2}{2}\right)}{\cos\left(\frac{\phi_1 - \phi_2}{2}\right)} \text{ and } y_1 = \frac{b \sin\left(\frac{\phi_1 + \phi_2}{2}\right)}{\cos\left(\frac{\phi_1 - \phi_2}{2}\right)} \qquad ...(3)$$

Now co-ordinates of P are $(a \cos \phi_1, b \sin \phi_1)$ and of the focus S are $(-ae, 0)$.

Hence the equation to PS is $y - 0 = \dfrac{b \sin \phi_1 - 0}{a \cos \phi_1 + ae}(x + ae)$

or $xb \sin \phi_1 - (a \cos \phi_1 + ae)\, y + abe \sin \phi_1 = 0$ ...(4)

If the line PS touches the circle whose centre is (x_1, y_1)

i.e., $$\left[\frac{a \cos \frac{(\phi_1 + \phi_2)}{2}}{\cos \frac{(\phi_1 - \phi_2)}{2}}, \frac{b \sin \frac{(\phi_1 + \phi_2)}{2}}{\cos \frac{(\phi_1 - \phi_2)}{2}}\right]$$

Then the length of perpendicular from (x_1, y_1) on (4) must be = radius of he circle. The length of the perpendicular

$$\frac{\dfrac{ab \sin \phi_1 \cos \frac{(\phi_1 + \phi_2)}{2}}{\cos\left(\frac{\phi_1 - \phi_2}{2}\right)} - ab(\cos \phi_1 + e)\dfrac{\sin \frac{\phi_1 + \phi_2}{2}}{\cos \frac{\phi_1 - \phi_2}{2}} + abe \sin \phi_1}{\left\{b^2 \sin^2 \phi_1 + a^2 (\cos \phi_1 + e)^2\right\}}$$

The numerator

$$\frac{ab}{\cos\left(\frac{\phi_1 - \phi_2}{2}\right)} \sin \phi_1 \cos \frac{\phi_1 + \phi_2}{2} - (\cos \phi_1 + e) \sin\left(\frac{\phi_1 + \phi_2}{2}\right)$$

$$+ e \sin \phi_1 \cos \frac{\phi_1 - \phi_2}{2}$$

$$\frac{a}{\cos\frac{\phi_1-\phi_2}{2}}\left[\left\{\sin\phi_1\cos\phi_1\sin\frac{\phi_1+\phi_2}{2}\right\}\right.$$

$$\left.+e\left\{\sin\phi_1\cos\phi_1\sin\frac{\phi_1-\phi_2}{2}-\sin\frac{\phi_1+\phi_2}{2}\right\}\right]$$

$$=\frac{ab}{\cos\frac{\phi_1-\phi_2}{2}}\left[\sin\frac{\phi_1-\phi_2}{2}+\frac{e}{2}\sin\left(\frac{\phi_1+\phi_2}{2}+\sin\frac{3\phi_1+\phi_2}{2}\right.\right.$$

$$\left.\left.-2\sin\frac{\phi_1+\phi_2}{2}\right)\right]$$

$$=\frac{ab}{\cos\frac{\phi_1-\phi_2}{2}}\left[\sin\frac{\phi_1-\phi_2}{2}+\frac{e}{2}\left(+\sin\frac{3\phi_1-\phi_2}{2}-\sin\frac{\phi_1+\phi_2}{2}\right)\right]$$

$$=\frac{ab}{\cos\frac{\phi_1-\phi_2}{2}}\left[\sin\frac{\phi_1-\phi_2}{2}+e\cos\phi_1\sin\frac{\phi_1-\phi_2}{2}\right]$$

$$=ab\tan\frac{\phi_1-\phi_2}{2}(1+e\cos\phi_1)$$

Denominator $= \sqrt{\{b^2\sin^2\phi_1 + a^2(\cos^2\phi_1 + e)^2\}}$

$= \sqrt{(b^2\sin^2\phi_1 + a^2\cos^2\phi_1 + a^2e^2 + 2a^2e\cos\phi_1)}$

$= \sqrt{\{a^2(1-e^2)\sin^2\phi_1 + a^2\cos^2\phi_1\ a^2e^2 + 2s^2e\cos\phi_1\}}$

[as $b^2 = a^2(1-e^2)$]

$\therefore$ Denominator $= \{a^2(1 + 2e\cos\phi_1 + e^2\cos^2\phi_1)\}$

$= \sqrt{\{a^2(1 + e\cos\phi_1)^2\}} = a(1 + e\cos\phi_1)$.

Hence $\dfrac{\text{numerator}}{\text{denominator}} = b\tan\dfrac{\phi_1-\phi_2}{2}$

Hence the length of perpendicular = b than $\left(\dfrac{\phi_1-\phi_2}{2}\right)$.

Similarly the lengths of the perpendicular from (x_1, y_1) on lines PS' QS and QS' are each equal to be tan $\dfrac{\phi_1-\phi_2}{2}$, the radius of the circle. Hence the above lines touch the circle drawn with the pole of PQ as centre and b tan $\{(\phi_1 - \phi_2)/2\}$ as radius.

Proved.

Example 61(c):

In the ellipse $\frac{x^2}{a^2} + \frac{y^2}{b^2} = 1$, *write down the equations to the diameters which are conjugate to the diameters whose equations are*

$x = y = 0,$

$x + y = 0,$

$y = \frac{a}{b}x$, and $y = \frac{b}{a}x$.

Solution:

As the equation to the ellipse is given as $x^2/a^2 + y^2/b^2 = 1$ and equation to one diameter is $x - y = 0$ (given). Hence its slope $m = 1$.

So the equation to the conjugate diameter will be

$$y = \frac{-b^2}{a^2 m}x = \frac{-b^2}{a^2}x \text{ (putting the value of m).}$$

Again the slope of the other diameter $x + y = 0$ is -1.

Hence the conjugate diameter is $y = \frac{-b^2}{a^2(-1)}x = \frac{b^2}{a^2}x$

And the slope of the third diameter $y = \frac{a}{b}x$ is $\frac{a}{b}$

Hence the conjugate diameter is $y = \frac{-b^2}{a^2}\left(\frac{b}{a}\right)x = \frac{-b^3}{a^2}x$

And the slope of the 4th diameter $y = \frac{b}{a}x$ is $\frac{b}{a}$

Hence the conjugate diameter is $y = \frac{-b^2}{a^2}.\frac{a}{b}x = -\frac{b}{a}x$ **Ans.**

Example 61(d):

In the question no. 70, If PN be the ordinate of P and the polar meet the axis in T, show that $CL = e^2.\ CN$ *and* $CT.CN = a^2$.

Solution:

The co-ordinates of L are

$$\left[x_1\left(1 - \frac{b^2}{a^2}\right), 0\right]$$

Hence $CL = x_1\left(1 - \frac{b^2}{a^2}\right) = x^{1}e^2$ [as $C \equiv (0, 0)$]

As $x_1 = CN$, so $CL = e^2.CN$.

Equation to the polar of P is $b^2xx_1 + a^2yy_1 = a^2b^2$.

The polar meets major-axis, i.e., x-axis in T where $y = 0$; so, solving with $y = 0$, we get $x = a^2/x_1 = CT$.

Hence $CT.CN = (a^2/x_1).\, x_1 = a^2$. **Proved.**

Example 62:

If the tangents TP and TQ be drawn from a point T whose coordinates are h and k, prove that the area of the triangle TPQ is

$$ab\left(\frac{h^2}{a^2} + \frac{k^2}{b^2} - 1\right)^{3/2} + \left(\frac{h^2}{a^2} + \frac{k^2}{b^2}\right),$$

and that the area of the quadrilateral CPTQ is (ab) $\left(\frac{h^2}{a^2} + \frac{k^2}{b^2} - 1\right)^{1/2}$

Solution:

If the co-ordinates of P and Q the points of contact on the ellipse

$\frac{x^2}{a^2} + \frac{y^2}{b^2} = 1$ be $(a\cos\phi_1, b\sin\phi_1)$ and $(a\cos\phi_2, b\sin\phi_2)$ respectively,

then

The equation of PQ will be

$$\frac{x}{a}\cos\frac{\phi_1 + \phi_2}{2} + \frac{y}{b}\sin\frac{\phi_1 + \phi_2}{2} = \cos\frac{\phi_1 - \phi_2}{2} \quad ...(1)$$

Again polar of the point T (h, k) w.r.t. the ellipse given will be

$$hx/a^2 + ky/b^2 = 1. \quad ...(2)$$

As the equations (1) and (2) represents the same line, therefore on comparing coefficient, we have

$$\frac{\frac{1}{a}\cos\frac{\phi_1 + \phi_2}{2}}{h/a^2} = \frac{\frac{1}{b}\sin\frac{\phi_1 + \phi_2}{2}}{k/b^2} = \cos\frac{\phi_1 - \phi_2}{2}$$

$$\Rightarrow \frac{\cos\frac{\phi_1 + \phi_2}{2}}{h/a} = \frac{\sin\frac{\phi_1 + \phi_2}{2}}{k/b} = \cos\frac{\phi_1 - \phi_2}{2} = \frac{1}{[(h/a)^2 + (k/b)^2]}$$

Hence $\cos\dfrac{\phi_1+\phi_2}{2} = \dfrac{h/a}{\sqrt{\left(\dfrac{b^2h^2+k^2a^2}{a^2b^2}\right)}} = \dfrac{bh}{\sqrt{(b^2h^2+a^2k^2)}}$

$\sin\dfrac{\phi_1+\phi_2}{2} = \dfrac{ak}{\sqrt{(b^2h^2+a^2k^2)}}$ and $\cos\dfrac{\phi_1-\phi_2}{2} = \dfrac{ab}{\sqrt{(b^2h^2+a^2k^2)}}$

Now, if TN and CN be the lengths of the perpendiculars from O and C on P, then

$$\frac{\Delta TPQ}{\Delta PCQ} = \frac{\frac{1}{2}\,PQ.TN}{\frac{1}{2}\,PQ.CM} = \frac{TN}{CM} \text{ or } \Delta PCQ \times \frac{TN}{CM}$$

Again $TN = \dfrac{\dfrac{h^2}{a^2}+\dfrac{k^2}{b^2}-1}{\sqrt{\left(\dfrac{h^2}{a^4}+\dfrac{k^2}{b^4}\right)}}$ and $CM = \dfrac{1}{\sqrt{\left(\dfrac{h^2}{a^4}+\dfrac{k^2}{b^4}\right)}}$

(neglecting negative sign)

$$\therefore \quad \frac{TN}{CN} = \frac{h^2}{a^2}+\frac{k^2}{b^2}-1.$$

As the area of the triangle having vertices as (x_1, y_1), (x_2, y_2) and (0, 0) is given by = ½ $(x_1x_2 - y_1y_2)$.

where (x_1, y_1) = the co-ordinates of P $(a\cos\phi_1, b\sin\phi_1)$

and (x_2, y_2) = the co-ordinates of Q $(a\cos\phi_2, b\sin\phi_2)$

$$\text{Hence } \Delta PCQ = \frac{1}{2}[ab\sin\phi_1\cos\phi_2 - ab\cos\phi_1\sin\phi_2]$$

$$= \frac{1}{2}ab\sin(\phi_1-\phi_2) = ab\sin\frac{\phi_1-\phi_2}{2}\cos\frac{\phi_1-\phi_2}{2}$$

$$= ab\cos\frac{\phi_1-\phi_2}{2}\sqrt{\left(1-\cos^2\frac{\phi_1-\phi_2}{2}\right)}$$

$$= ab\frac{ab}{\sqrt{(b^2h^2+a^2k^2)}}\sqrt{\left(1-\frac{a^2b^2}{(b^2h^2+a^2k^2)}\right)}$$

$$= ab \frac{1}{\sqrt{(h^2/a^2 + k^2/b^2)}} \times \sqrt{\left(1 - \frac{1}{(h^2/a^2 + k^2/b^2)}\right)}$$

$$= \frac{ab\sqrt{[h^2/a^2 + k^2/b^2 - 1]}}{(h^2/a^2 + k^2/b^2)}$$

Hence the $\Delta TPQ = \Delta PCQ \times \frac{TN}{CM}$

$$= \frac{ab\sqrt{[h^2/a^2 + k^2/b^2 - 1]}}{(h^2/a^2 + k^2/b^2)} (h^2/a^2 + k^2/b^2 - 1)$$

$$= \frac{ab\,[h^2/a^2 + k^2/b^2 - 1]^{3/2}}{(h^2/a^2 + k^2/b^2)}.$$ **Proved.**

Again area of the quadrilateral CPTQ = DTPQ + DPCQ

$$= \frac{ab\,[h^2/a^2 + k^2/b^2 - 1]^{3/2}}{h^2/a^2 + k^2/b^2} + \frac{ab\,[h^2/a^2 + k^2/b^2 - 1 + 1]^{1/2}}{h^2/a^2 + k^2/b^2}$$

Example 63:

CK is the perpendicular from the centre on the polar of any point P, and PM is the perpendicular from P on the same polar and is produced to meet the major axis in L, show that (1) CK.PL = b^2 and (2) the product of the perpendiculars from the foci on the polar = CK.LM.

What do these theorems become when P is on the ellipse.

Solution:

Let p be any point (x_1, y_1) and the equation to the ellipse be

$$\frac{x^2}{a^2} + \frac{y^2}{b^2} = 1. \qquad \ldots(1)$$

The equation to polar of P w.r.t. (1) is $\frac{xx_1}{a^2} + \frac{yy_1}{b^2} = 1$

$$\Rightarrow \quad b^2xx_1 + a^2yy_1 = a^2b^2 \qquad \ldots(2)$$

Length of the perpendicular CK from C = (0, 0) on (2) we have

$$= \frac{a^2b^2}{\sqrt{(b^4x_1^2 + a^4y_1^2)}}$$

Also length of the perpendicular from P on the polar (2)

$$= PM = \frac{a^2x_1^2 + a^2y_1^2 - a^2b^2}{\sqrt{(b^4x_1^2 + a^4y_1^2)}}$$

Again the slope of the polar of P, i.e., (2) $= -\frac{b^2}{a^2} \times \frac{x_1}{y_1}$

$\therefore$ Slope of PM $= \frac{a^2}{b^2} \times \frac{y_1}{x_1}$ [as PM is perpendicular to (2)]

So the equation of PM will be $a^2y_1(x - x_1) - b^2x_1(y - y_1) = 0$.

If this meets major axis at point L, then solving with x-axis, i.e., y = 0,

we have $x = x_1\left(1 - \frac{b^2}{a^2}\right)$.

Hence the co-ordinates of L are given as $= \left[x_1\left(1 - \frac{b^2}{a^2}\right), 0\right]$

$$\therefore \quad PL = \sqrt{\left[\left\{x_1\left(1 - \frac{b^2}{a^2}\right) - x_1\right\}^2 + y_1^2\right]} = \sqrt{\left(x_1^2\frac{b^2}{a^4} + y_1^2\right)}$$

$$= \frac{1}{a^2}\sqrt{(b^4x_1^2 + a^4y_1^2)}.$$

$$\text{Now CK.PL} = \frac{a^2b^2}{\sqrt{(b^4x_1^2 + a^4y_1^2)}} \cdot \frac{\sqrt{(b^4x_1^2 + a^4y_1^2)}}{a^2} = b^2$$

$$\text{and LM} = PL - PM = \frac{\sqrt{(b^4x_1^2 + a^4y_1^2)}}{a^2} - \frac{b^4x_1^2 + a^2y_1^2 - a^2b^2}{\sqrt{(b^4x_1^2 + a^4y_1^2)}}$$

$$= \frac{(b^4x_1^2 + a^4y_1^2x_1^2 - a^4y_1^2 + a^4b^2)}{a^2\sqrt{(b^4x_1^2 + a^4y_1^2)}}$$

$$\Rightarrow \quad LM = \frac{b^2(b^2x_1^2 - a^2x_1^2 + a^4)}{a^2\sqrt{(b^4x_1^2 + a^4y_1^2)}} = \frac{b^2(a^2 - e^2x_1^2)}{\sqrt{(b^4x_1^2 + a^4y_1^2)}}$$

[as $b^2 = a^2(1 - e^2)$]

Again the product of lengths of perpendicular from foci (ae, 0) and (–ae, 0) on the polar (2)

$$= \frac{b^2aex_1 - a^2b^2}{a^4\sqrt{(b^4x_1^2 + a^4y_1^2)}} \times \frac{-b^2aex_1 - a^2b^2}{(b^4x_1^2 + a^4y_1^2)}$$

$$= \frac{a^4b^4 - b^4a^2e^2x_1^2}{b^4x_1^2 + a^4y_1^2} = \frac{b^4a^2(a^2 - e^2x_1^2)}{b^4x_1^2 + a^4y_1^2}$$

Also CK.LM $= \dfrac{b^4a^2(a^2 - e^2x_1^2)}{b^4x_1^2 + a^4y_1^2}$ = product of the lengths of the perpendiculars from the foci in polar.

If P is on the ellipse, P and M will coincide and therefore PK will be a tangent at P and PL normal. **Proved.**

Example 64:

Tangents are drawn to the ellipse from the point

$$\left(\frac{a^2}{\sqrt{a^2 - b^2}}, \sqrt{a^2 + b^2}\right),$$

prove that they intersection the ordinate through the nearer focus a distance equal to the major axis.

Solution:

The equation to the ellipse is $x^2/a^2 + y^2/b^2 = 1$.

The co-ordinates of the point are $\left\{\dfrac{a^2}{\sqrt{a^2 - b^2}}, \sqrt{(a^2 + b^2)}\right\}$

$\Rightarrow \quad \{a/e, \sqrt{(a^2 + b^2)}\}$.

The equation to the pair of tangents will be given by $SS_1 = T^2$

where $S = (x^2/a^2 + y^2/b^2 - 1)$

$$S_1 = \frac{a^2}{c^2a^2} + \frac{a^2 + b^2}{b^2} - 1 = \frac{1}{e^2} + \frac{a^2}{b^2} = \frac{1}{e^2} + \frac{1}{1 - e^2} = \frac{1}{e^2(1 - e^2)}$$

and $\quad T = \dfrac{x}{ae} + \dfrac{y\sqrt{(a^2 + b^2)}}{b^2} - 1.$

{i.e., the equation to the tangent from {a/e, $\sqrt{(a^2 + b^2)}$}, as if this point is on the ellipse}.

So equation to the pair of tangents will be

$$\frac{1}{e^2(1-e^2)}\left(\frac{x^2}{a^2}+\frac{y^2}{b^2}-1\right)=\left(\frac{x}{ae}+\frac{y\sqrt{(a^2+b^2)}}{b^2}-1\right)^2$$

This tangent meets the latus rectum in the point where x = ae, so putting the value of x, we get

$$\frac{1}{-e^2}+\frac{y^2}{e^2 b^2(1-e^2)}=\frac{y^2(a^2+b^2)}{b^2} \text{ and as } b^2 = a^2(1-e^2)$$

we have as $\dfrac{1}{-e^2}+\dfrac{y^2}{a^2 e^2(1-e^2)}=\dfrac{y^2(2a^2-a^2e^2)}{a^4(1-e^2)}$

$$\Rightarrow \quad y^2\left[\frac{1}{a^2 e^2(1-e^2)}-\frac{2-e^2}{a^2(1-e^2)^2}\right]=\frac{1}{e^2}$$

or $\dfrac{y^2(1-2e^2+e^4)}{a^2 e^2(1-e^2)}\ \dfrac{1}{e^2}$ or $\dfrac{y^2(1-e^2)^2}{a^2e^2(1-e^2)^2}=\dfrac{1}{e^2}$

Let y_1 and y_2 be the ordinates of the points; then

we have $y_1 - y_2 = a - (-a) = 2a$ = major axis. **Proved.**

Example 65(a):

Show that the diameters whose equations are $y + 3x = 0$ and $4y - x = 0$, are conjugate diameters of the ellipse $3x^2 + 4y^2 = 5$.

Solution:

The equation to the ellipse is given as $3x^2 + 4y^2 = 5$

$$\Rightarrow \quad \frac{x^2}{5/3}+\frac{y^2}{5/4}=1$$

Hence we have $a^2 = \dfrac{5}{3}$ and $b^2 = \dfrac{5}{4}$

The slope of one diameter, $y + 3x = 0$ (given is $-3 = m_1$ (say).

The conjugate diameter will be $y = \dfrac{b^2\,x}{a^2\,m} = -\dfrac{5/4}{5/3}\times = -\dfrac{1}{4}x$

$\Rightarrow$ $4y + x = -0$, which is the other given diameter. **Proved.**

Example 65(b):

If the product of the perpendiculars from the foci upon the polar of P be constant and equal to c^2, prove that the locus of P is the ellipse

$$b^4x^2 (c^2 + a^2e^2) + c^2a^2y^4 = a^4 b^4.$$

Solution:

Suppose the equation tot he ellipse is $x^2/a^2 + y^2/b^2 = 2$.

The co-ordinates of foci are (ae, 0) and (–ae, 0).

Let the co-ordinates of P be (h, k). Then polar of P is

$$\frac{xh}{a^2} = -\frac{yk}{b^2} = 1 \text{ or } b^2 xh + a^2 yk - a^2 b^2 = 0$$

If p^1 and p^2 be the lengths of the perpendiculars on the line (1) from (ae, 0) and (–ae, 0) on (1), respectively.

$$\text{then } p_1 = \frac{b^2 hae - a^2 b^2}{\sqrt{(b^4 h^2 + a^4 k^2)}} \text{ and } p_2 = \frac{-a^2b^2 - b^2 hea}{\sqrt{(b^4 h^2 + a^4 k^2)}}$$

$$\therefore \quad p_1 p_2 = \frac{-b^4 h^2 a^2 e^2 + a^4 b^4}{b^4 h^2 + a^4 k^4} = c^2 \quad \text{(by hypothesis)}$$

$$\Rightarrow \quad a^4b^4 = b^4h^2a^2e^2 + c^2b^4h^2 + c^2a^4k^2$$

$$\Rightarrow \quad b^4h^2 (c^2 + a^2e^2) + c^2a^4k^2 = a^4b^4.$$

Generalising the locus of the point P (h, k) is

$$b^4x^2 (c^2 + a^2e^2) + c^2a^4y^2 = a^4b^4.$$ **Proved.**

Example 66:

If T be the point (x_1, y_1), show that the equation to the straight lines joining it to the foci, S and S', is $(x_1y - xy_1)^2 = a^2e^2 (y - y_1)^2 = 0$.

Prove that the bisector of the angle between these lines also bisects the angle between the tangents TP and TQ that can be drawn from T and hence that ∠STP = ∠STQ.

Solution:

If the equation to the ellipse be $x^2/a^2 + y^2/b^2 = 1$, then co-ordinates of the foci S and S' are (–ae, 0) and (ae, 0) ;respectively and of T are given as (x_1, y_1).

$$\text{Then the equation to Ts is } y = 0 = \frac{y_1 - 0}{x_1 + ae}(x + ae)$$

$\Rightarrow \qquad (x_1y - xy_1) + ae\ (y - y_1) = 0.$

The equation to TS' is in the same way

$$(x_1y - xy_1) - ae\ (y - y_1) = 0.$$

The combined equation of TS and TS' will be

$\{(x_1y - xy_1) + ae\ (y - y_1)\}\ \{(x_1y - xy_1) - ae\ (y - y_1) = 0$

$\Rightarrow \quad (x_1y - xy_1)^2 - a^2e^2\ (y - y_1)^2 = 0.$ ...(1) **Proved.**

Now; the equation of pair of lines parallel to the lines given by (1), passing through the origin, is $y_1{}^2x^2 - 2x_1y_1xy + y^2\ (x_1{}^2 - a^2e^2) = 0.$

Hence the equation to the bisector of the angle between the lines is

$$\frac{x^2 - y^2}{y_1^2 - \left(x_1^2 - a^2\ e^2\right)} = \frac{xy}{-x_1y_1}$$

$$\Rightarrow \quad \frac{x^2 - y^2}{y_1^2 - x_1^2 + a^2 - b^2} = \frac{xy}{-x_1y_1} \left(\text{as } e^2 = \frac{a^2 - b^2}{a^2}\right) \quad ...(2)$$

We can prove that the equation of the pair of lines parallel to tangents TP and TQ from T, passing through the origin is $x\ (y_1{}^2 - b^2) - 2xyx_1y_1 + y^2\ (x_1{}^2 - a^2) = 0.$

Equation to the bisector of these lines will be

$$\frac{x^2 - y^2}{\left(y_1^2 - b^2\right)\left(x_1^2 - a^2\right)} = \frac{xy}{-x_1y_1}$$

$$\Rightarrow \quad \frac{x^2 - y^2}{y_1^2 - x_1^2 - a^2 - b^2} = \frac{xy}{-x_1y_1} \quad ...(3)$$

As (2) and (3) are same, hence it follows that the bisectors of angle STS' is the same as that of the angle

PTQ, So ∠ STP = ∠ STQ.

Example 67(a):

If two tangents to an ellipse and one of its foci be given, prove that the locus of its centre is a straight line.

Solution:

Suppose TP and TQ be two tangents from T, on the ellipse having the focus at S. As ∠ QTS' = ∠ PTS (by Q. 74), the centre C of the ellipse is mid-way between S and S', hence when S' moves on TS', C will move parallel to TS' so that it bisects the line SS'. Therefore the locus of centre C is a straight line. **Proved.**

Example 67(b):

Prove that the straight lines joining the centre to the intersections of the straight line $y = mx + \sqrt{(a^2m^2 + b^2)/2}$, *with the ellipse are conjugate diameters.*

Solution:

Let the equation to the ellipse be

$$x^2/a^2 + y^2/b^2 = 1 \quad ...(1)$$

and equation to the line is given as $y = mx + \sqrt{\left(\frac{a^2m^2b^2}{2}\right)}$

$$\Rightarrow \quad \frac{(y-mx)\sqrt{2}}{\sqrt{(a^2m^2+b^2)}} = 1. \quad ...(2)$$

To get the equation to the line joining the point of intersection of the origin, making (1) homogenous we are help of (2), we have

$$\frac{x^2}{a^2} + \frac{y^2}{b^2} = \left[\frac{(y-mx)\sqrt{2}}{\sqrt{(a^2m^2+b^2)}}\right]^2 = \frac{2(y-m^2x^2-2mxy)}{(a^2m^2+b^2)}$$

$$\Rightarrow \quad (b^2x^2 + a^2y^2)(a^2m^2 + b^2) = 2a^2b^2(y^2 + m^2x^2 - 2mxy)$$

$$\Rightarrow \quad y^2a^2(a^2m^2 + b^2) + 4ma^2b^2xy - b^2x^2 + m^2x^2(a^2m^2 - b^2) = 0$$

$$\Rightarrow \quad y^2 + \frac{4mb^2}{a^2m^2-b^2}xy - \frac{b^2}{a^2}x^2 = 0$$

This equation represents two straight lines $y = M_1x$ and $y = M_2x$ then be combined equation will be

$$y^2 - (M_1 + M_2)yx + M_1M_2x^2 = 0. \quad ...(4)$$

Comparing (3) and (4), we get $M_1M_2 = \frac{b^2}{a^2}$

which is the condition of diameters to be conjugate.

Hence the lines are the conjugate diameters. **Proved.**

Example 67(c):

Prove that the angle between the tangents that can be drawn from any point (x_1, x_2) *to the ellipse is*

$$\tan^{-1} = \frac{2ab\sqrt{\frac{x_1^2}{a^2} + \frac{y_1^2}{b^2} - 1}}{x_1^2 + y_1^2 - a^2 - b^2}$$

Solution:

As equation to the pair of tangents drawn from any point (x_1, y_1) on the ellipse $x^2/a^2 + y^2/b^2 = 1$ is given by $SS_1 = T^2$, ...(1)

where $S \equiv x^2/a^2 + y^2/b^2 - 1,$

$S_1 \equiv x_1^2/a^2 + y_1^2/b^2 - 1.$

and $T \equiv xx_1/a^2 + yy_1/b^2 - 1.$

Putting the value, in (1), the equation to the pair of tangent will be $(x^2/a^2 + y^2/b^2 - 1)(x_1^2/a^2 + y_1^2/b^2 - 1) \equiv (xx_1/a^2 + yy_1/b^2 - 1)^2.$

$$\frac{x^2}{a^2}\left(\frac{x_1^2}{b^2} + \frac{y_1^2}{a^2} - 1 - \frac{x_1^2}{a^2}\right) - \frac{2xy\,x_1\,y_1}{a^2\,b^2} + \frac{y^2}{b^2}\left(\frac{x_1^2}{a^2} + \frac{y_1^2}{b^2} - 1 - \frac{y_1^2}{b^2}\right)$$

$$+ Ax + By + C = 0$$

where A and B are the coefficients of x and y and C is the constant term on simplification

$\Rightarrow x^2(y_1^2 - b^2) - 2xyx_1y_1 + y^2(x_1^2 - a^2) +$ terms of lower degree $= 0$ ...(2)

If the angle between the tangents be θ, then tan θ is given by

$$\tan\theta = \frac{2\sqrt{(h^2 - ab)}}{a + b} = \frac{2\sqrt{\left\{x_1^2 y_1^2 - \left(y_1^2 - b^2\right)\left(x_1^2 - a^2\right)\right\}}}{\left(y_1^2 - b^2\right) + \left(x_1^2 - a^2\right)}$$

$$\text{or } \theta \tan^{-1}\frac{2\sqrt{\left(b^2\,x_1^2 + a^2\,y_1^2\right) - a^2\,b^2}}{x_1^2 + y_1^2 - a^2 - b^2} = \tan^{-1}\frac{2ab\sqrt{\left(\frac{x_1^2}{a^2} + \frac{y_1^2}{b^2} - 1\right)}}{x_1^2 + y_1^2 - a^2 - b^2}$$

Example 68:

A pair of conjugate diameters is produced to meet the directrix; show that the orthocentre of the triangle so formed is at the focus.

Solution:

If the co-ordinates of P are taken as (a cos φ, b sin φ), then the co-ordinates of Q' will be

(a cos (φ + π/2), b sin (φ + π/2)} i.e., (–a sin φ, b cos φ), as these are the ends of conjugate diameters CP and Q'C of the ellipse $\frac{x^2}{a^2} + \frac{y^2}{b^2} = 1.$

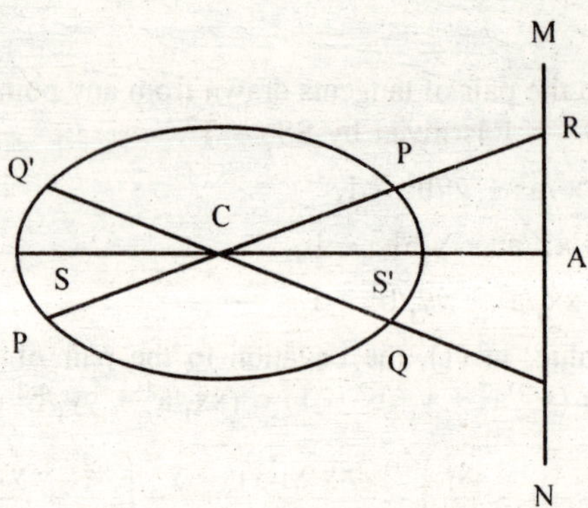

If these diameters are produced, let them meet the directrix in points R and S; then we have to find the orthocentre of the so formed triangle CRS.

Clearly, the equation to P'CP will be $y = \frac{bx}{a} \tan \phi_1$ ...(1)

and that of Q'CQ will be

$$y = \frac{-bx}{a} \cot \phi \qquad ...(2).$$

On solving (1) and (2) with directrix x = a/e on by one, the co-ordinate of R and S are found to be $\left(\frac{a}{e}, \frac{b}{e} \tan \phi\right)$ and $\left(\frac{a}{e}, \frac{-b}{e} \cot \phi\right)$ respectively.

Again the equation to be perpendicular from R on CS is

$$\left(y - \frac{b}{e} \tan \phi\right) = \frac{a}{b} \tan \phi \left(x - \frac{a}{e}\right) \qquad ...(3)$$

The line perpendicular from the centre C upon the directrix MN will be x-axis, i.e., y = 0.

Solving (3) with y = 0, we have $\frac{-b}{a} \tan \phi = \frac{a}{b} \tan \phi = \left(x - \frac{a}{e}\right)$ whence,

$$x = \frac{a}{e} - \frac{b^2}{ae} = \frac{a^2 - b^2}{ae} = \frac{a^2 e^2}{ae} = ae.$$

So the perpendicular meet at (ae, 0), therefore the co-ordinates of orthocentre are (ae, 0) which are the co-ordinates of focus.

Example 69:

Any tangent to an ellipse meets the area on circle in P and d; prove that CP and CD are in the direction of conjugate diameters of the ellipse.

Solution:

If the equation to the ellipse be $x^2/a^2 + y^2/b^2 = 1$; then $y = mx + \sqrt{(a^2m^2 + b^2)}$ ...(1) is always tangent on it

The equation to the director circle is $x^2 + y^2 = a^2 + b^2$. ...(2)

The equations to the lines Cp and Cd the lines joining the points of intersection of (1) with (2) are obtained by making (2) homogeneous with the help of (1) so, we get

$$x^2 + y^2 = \left(a^2 + b^2\right)\left\{\frac{y - mx}{\sqrt{\left(a^2 m^2 + b^2\right)}}\right\}^2$$

$\Rightarrow$ $(x^2 + y^2)(a^2m^2 + b^2) = (a^2 + b^2)(y^2 - 2mxy + m^2x^2)$

$\Rightarrow$ $y^2a^2(m^2 - 1) + 2mxy(a^2 + b^2) + x^2b^2(1 - m^2) = 0.$ ...(3)

As in the last question, if $y = M_1x$ and $y = M_2x$ be the lines represented by (3), we have

$$M_1 M_2 = \frac{b^2(1 - m^2)}{a^2(m^2 - 1)} = -\frac{b^2}{a^2}.$$

which is again the condition for the diameters to be conjugate.

Therefore CP and CD are the conjugate diameters. **Proved.**

Example 70:

If CP be conjugate to the normal at Q, prove that CQ is conjugate to the normal at P.

Solution:

If the co-ordinates of P and Q be taken as $(a \cos\alpha, b \sin\alpha)$ respectively, then the equation to the normal at Q will be $ax \sec\beta \operatorname{cosec}\beta = a^2 - b^2$.

Slope of this normal $= \dfrac{a \sec\beta}{a \operatorname{cosec}\beta} = \dfrac{a}{b}\tan\beta = m_1$ (say)

Also the slope of CP $= \dfrac{b \sin\alpha}{a \cos\alpha} = \dfrac{b}{a}\tan\alpha = m_2$ (say)

If CP is the conjugate diameters to the normal at Q then we must have

$$m_1m_2 = b^2/a^2 \text{ or } (a/b)\tan\beta.(b/a)\tan\alpha = -b^2/a^2.$$

Hence $\tan\alpha \tan\beta = -b^2/a^2$.

As this relation is symmetrical in α and β, it proves that CQ is also a conjugate diameter to normal at P. **Proved.**

Example 71:

If circles be described on two semi-conjugate diameters of the ellipse as diameters, prove that the locus of their second points of intersection is the curve

$$2(x^2 + y^2)^2 = a^2x^2 + b^2y^2.$$

Solution:

If CP and CQ be the two conjugate diameters of the ellipse $x^2/a^2 + y^2/b^2 = 1$, having the centre $C \equiv (0, 0)$ and the co-ordinates of P be $(a \cos \phi, b \sin \phi)$, then the co-ordinates of Q are $(-a \sin \phi, b \cos \phi)$. The equation to the circles drawn with CP and CQ as diameters are respectively

$$(x - 0)(x - a \cos \phi) + (y - 0)(y - b \sin \phi) = 0$$

$$\Rightarrow \quad x^2 + y^2 = ax \cos \phi - by \sin \phi = 0 \qquad ...(1)$$

$$\text{and} \quad (x - 0)(x + a \sin \phi) + (y - 0)(y - b \sin \cos \phi) = 0$$

$$\Rightarrow \quad x^2 + y^2 - (ax \sin \phi + by \cos \phi) = 0. \qquad ...(2)$$

To find the locus of point of intersection of (1) and (2), we have to eliminate the variable ϕ from them. Subtracting (1) from (2), we get

$$(ax + by) \sin \phi = (by = ax) \cos \phi$$

$$\Rightarrow \quad \frac{\sin \phi}{by - ax} = \frac{\cos \phi}{by + ax} = \frac{1}{\sqrt{[(by - ax)^2 + (by + ax)^2]}}$$

$$= \frac{1}{\sqrt{[2(b^2y^2 + a^2x^2)]}}$$

$$\text{Hence } \sin \phi = \frac{by - ax}{\sqrt{[2(b^2y^2 + a^2x^2)]}} \text{ and } \cos\phi = \frac{by + ax}{\sqrt{2(b^2y^2 + a^2x^2)}}$$

Now putting the values for $\sin \phi$ and $\cos \phi$ in (1), we get

$$x^2 + y^2 = \frac{ax(by + ax)}{\sqrt{[2(b^2y^2 + a^2x^2)]}} + \frac{by(by - ax)}{\sqrt{2(b^2y^2 + a^2x^2)}}$$

$$= \frac{a^2x^2 + a^2b^2}{\sqrt{[2(a^2x^2 + b^2y^2)]}} = \frac{(a^2x^2 + b^2y^2)}{\sqrt{2}}$$

Squaring and simplifying the required locus of point of intersection is

$$2(x^2 + y^2) = (a^2x^2 + b^2y^2).$$

Example 72:

If the tangent at any point P meet in the points L and L' (1) two parallel tangents, or (2) two conjugate diameters, prove that in each case the rectangle LP.PL' is equal to the square on the semidiameter which is parallel to the tangent at P.

Solution:

(a) Let the conjugate diameters CP and CQ be the axis of x and axis of y respectively; then the equation to the ellipse is

$$\frac{x^2}{a^2}+\frac{y^2}{b^2}=1,$$

the co-ordinates of P and Q will be (a, 0) and (0, b) respectively. Then the equation to parallel tangent at (x_1, y_1) is given by

$$\frac{xx_1}{a^2}+\frac{yy_1}{b^2}=\pm 1. \qquad ...(1)$$

Also the tangent at P is $x - a = 0$. ...(2)

On solving (1) and (2), we get $y=\frac{b^2}{y_1}\left(-\frac{x_1}{a}\pm 1\right)$ (omitting –ve sign).

So LP So $LP=\frac{b^2}{y_1}\left(1-\frac{x_1}{a}\right)$ and $L'P=\frac{b^4}{y_1^2}\cdot\frac{y_1^2}{b^2}=b^2=CQ^2$

$$\left[\text{as } (x_1, y_1) \text{ lie on the ellipse } \left(\frac{x_1^2}{a^2}+\frac{y_1^2}{b^2}=1\right)\right]$$

(b) Any two conjugate diameters are given by

$$y=\frac{b}{a}x\tan\phi \text{ and } y=\frac{-b}{a}c\cot\phi$$

Solving these with tangent x = a, at point P, one by one, we get

$$y = b\tan\phi \text{ and } y = -b\cot\phi$$

So $LP = b\tan\phi$ and $L'P = -b\cot\phi$.

LP L'P = b^2 (numerically) or CQ^2. **Proved.**

Example 73:

CP and CD are conjugate diameters of the ellipse, prove that the locus of the orthocentre of the triangle CPD is the curve

$$2(b^2y^2 + a^2x^2)^3 = (a^2 - b^2)^2\,(b^2y^2 - a^2x^2)^2.$$

Solution:

If the co-ordinates of P are (a cos ϕ, b sin ϕ), then the co-ordinates of D will be $\left[a\cos\left(\frac{\pi}{2}+\phi\right), b\sin\left(\frac{\pi}{2}+\phi\right)\right]$ i.e, (–a sin ϕ, b cos ϕ), where CP and CD are the conjugate diameters of the ellipse $x^2/a^2 + y^2/b^2 = 1$, having centre as C(0, 0).

The equation to CP will be $y = \frac{b}{a}x = \tan\phi$.

and the equation to CD will be $y = \frac{-b}{a}x = \cot\phi$.

Now the equation of the line through D and perpendicular to CP will be

be $\quad y - b\cos\phi = \frac{-a}{b}\cot\phi\,(x + a\sin\phi)$

$\Rightarrow \quad ax\cos\phi + by\sin\phi = -(a^2 - b^2)\sin\phi\cos\phi.$ ...(1)

Similarly the equation to the line through P and perpendicular on CD is $\quad y - b\sin\phi = (a/b)\tan\phi\,(x - a\cos\phi)$

$\Rightarrow$ $ax\sin\phi - by\cos\phi = (a^2 - b^2)\sin\phi\cos\phi.$ ...(2)

The locus of the point of intersection of (1) and (2) will be the locus of the orthocentre of the triangle CPD. Do, we have to eliminate the variable from (1) and (2).

Adding (1) and (2), we get $(ax + by)\sin\phi + (ax - by)\cos\phi = 0$

$\Rightarrow \quad (ax + by)\sin\phi = (by - ax)\cos\phi$

Hence

$$\frac{\sin\phi}{by - ax} = \frac{\cos\phi}{by + ax} = \frac{1}{\sqrt{[(by - ax)^2 + (by + ax)^2]}} = \frac{1}{\sqrt{[2b^2y^2 + a^2x^2]}}$$

Hence $\sin\phi = \dfrac{by - ax}{\sqrt{2(b^2y^2 + a^2x^2)}}$

and $\quad \cos\phi = \dfrac{by + ax}{\sqrt{[2(b^2y^2 + a^2x^2)]}}$

Substituting the values of sin ϕ and cos ϕ in (1), we get

$$\frac{ax(by + ax)}{\sqrt{[2(a^2x^2 + b^2y^2)]}} + \frac{by(by - ax)}{\sqrt{[2(a^2x^2 + b^2y^2)]}}$$

$$= \frac{-(a^2 - b^2)(b^2y^2 - a^2x^2)}{2(a^2x^2 + b^2y^2)}$$

$$\Rightarrow \quad \sqrt{a^2x^2 + b^2y^2} = \frac{-(a^2 - b^2)(b^2y^2 - a^2x^2)}{2(a^2x^2 + b^2y^2)}$$

On squaring and corresponding we get the required locus as

$$2(a^2x^2 + b^2y^2)^3 = (a^2 - b^2)^2 (b^2y^2 - a^2x^2)^2.$$

Example 74:

A point is such that the perpendicular from the centre on its polar with respect to the ellipse is constant and equal to c; show that its locus is the ellipse $x^2/a^4 + y^2/b^4 = 1/c^2$.

Solution:

Let the co-ordinates of any point be (h, k) and equation to ellipse be $x^2/a^2 + y^2/b^2 = 1$. ...(1)

Now the polar of (h, k) w.r.t. (1) is $xh/a^2 + yk/b^2 = 1$ ...(2)

The length of perpendicular from the centre (0, 0) to (2) is given by

$$\frac{1}{\sqrt{(h^2/a^2 + k^2/b^2)}} = c \text{ (by hypothesis);}$$

on squaring and cross multiplying, we get $\frac{1}{2} = \frac{h^2}{a^4} + \frac{k^2}{b^4}$.

Generalising the locus of (h, k), we get $(x^2/a^4 + y^2/b^4) = 1/c^2$. **Proved.**

Example 75:

Tangents are drawn from any point on the ellipse $x^2/a^2 + y^2/b^2 = 1$, to the circle $x^2 + y^2 = r^2$; prove that the chords of contact are tangents to the ellipse $a^2x^2 + b^2y^2 = r^4$.

Solution:

(a) If the co-ordinates of any point on the ellipse $x^2/a^2 + y^2/b^2 = 1$ be (it cos φ, b sin φ; then the equation to its chord of contact with respect to the given circle $x^2 + y^2 = r^2$ is $ax \cos\phi + by \sin\phi = r^2$

$$\Rightarrow \quad \frac{x}{(r^2/a)} \cos\phi + \frac{y}{(r^2/b)} \sin\phi = 1. \qquad ...(1)$$

Equation (1) always touches the ellipse given by the equation

$$\frac{x^2}{(r^2/a)} + \frac{y^2}{(r^2/b)} = 1$$ **Proved.**

(b) The combined equation to the lines joining the points of intersection of circle $x^2 + y^2 = r^2$ with the line $ax \cos \phi + by \sin \phi = r^2$ is obtained by making the equation of circle homogeneous with the help of the line, i.e.,

$$x^2 + y^2 = r^2 \frac{(ax \cos \phi + by \sin \phi)^2}{(r^2)^2}$$

$$\Rightarrow \quad r^2 (x^2 + y^2) = a^2x^2 \cos^2 \phi + b^2y^2 \sin^2 \phi + 2abxy \sin \phi \cos \phi$$

$$\Rightarrow \quad x^2 (r^2 - a^2 \cos^2 \phi) - 2abxy \sin \phi \cos \phi$$

$$+ y^2 (r^2 - b^2 \sin^2 \phi) = 0 \qquad ...(1)$$

If the two lines given (1) are the conjugate diameters of the ellipse $a^2x^2 + b^2y^2 = r^4$, the required condition is

$$m_1m_2 = \frac{-a^2}{b^2} \text{ or } \frac{r^2 - a^2 \cos^2 \phi}{r^2 - b^2 \sin^2 \phi} = -\frac{a^2}{b^2}$$

$$\Rightarrow \quad b^2r^2 - a^2b^2 \cos^2 \phi = -a^2r^2 + a^2b^2 \sin^2 \phi$$

$$\Rightarrow \quad r^2 (a^2 + b^2) = a^2b^2 (\sin^2 \phi \cos^2 \phi) = a^2b^2$$

$$\Rightarrow \quad \frac{1}{r^2} = \frac{a^2 + b^2}{a^2b^2} = \frac{1}{a^2} + \frac{1}{b^2}$$ **Proved.**

Example 76:

Find the inclination to the major axis of the diameter of the ellipse the square of whose length is (1) the arithmetical mean. (2) the geometrical mean, and (3) the harmonical mean, between the squares on the major and minor axes.

Solution:

The polar equation to the ellipse

$$\frac{x^2}{a^2} + \frac{y^2}{b^2} = 1 \text{ is } \frac{r^2 \cos^2 \theta}{a^2} + \frac{r^2 \sin^2 \theta}{b^2} = 1$$

$$\Rightarrow \quad r^2 = \frac{a^2 b^2}{a^2 \sin^2 \theta + b^2 \cos^2 \theta}$$

(a) Again $(2r)^2 = \frac{(2a)^2 + (2b)^2}{2}$ [by hypothesis]

$$\Rightarrow \quad r^2 = \frac{2a^2\ b^2}{a^2\ \sin^2\theta + b^2\ \cos^2\theta} = a^2 + b^2$$

$$\Rightarrow \quad 2a^2b^2 = (a^2 + b^2)\ (a^2 \sin^2\theta + b^2 \cos^2\theta)$$

$$\Rightarrow \quad 2a^2b^2 \sec^2\theta = (a^2 + b^2)\ (a^2 \tan^2\theta + b^2)$$

[dividing both sides by $\cos^2\theta$]

$$\Rightarrow \quad 2a^2b^2\ (1 + \tan^2\theta) = (a^2 + b^2)\ (a^2 \tan^2\theta + b^2)$$

$$\Rightarrow \quad \tan^2\theta = (2a^2b^2 - a^4 - a^2b^2) = a^2b^2 + b^4 - 2a^2b^2$$

$$\tan^2\theta = \frac{b^2\ (b^2 - a^2)}{a^2\ (b^2 - a^2)}$$

$$\Rightarrow \quad \tan\theta = b/a, \text{ hence } \theta \tan^2 (b/a)$$ **Ans.**

(b) If $(2r)^2 = \sqrt{[2a)^2. (2b)^2]}$, then $r^2 = ab$;

$$\text{So} \quad \frac{a^2\ b^2}{a^2\ \sin^2\theta + b^2\ \cos^2\theta} = ab$$

$$\Rightarrow \quad \frac{ab\ (\sin^2\theta + \cos^2\theta)}{a^2\ \sin^2\theta + b^2\ \cos^2\theta} = 1 \qquad [\because \sin^2\theta + \cos^2\theta = 1]$$

$$\Rightarrow \quad ab \sin^2\theta + ab \cos^2\theta = a^4 \sin^2\theta + b^2 \cos^2\theta$$

$$\Rightarrow \quad (ab - a^2) \sin^2\theta = (b^2 - ab) \cos^2\theta$$

$$\Rightarrow \quad \tan^2\theta\ \frac{b^2 - ab}{ab - a^2} = \frac{b(b - a)}{a\ (b - a)} = \frac{b}{a}$$ **Ans.**

$$\Rightarrow \quad \tan\theta = \sqrt{(b/a)} \qquad \therefore \quad \theta = \tan^{-1} \sqrt{(b/a)}.$$

(c) If $\frac{2}{(2r)^2} = \frac{1}{(2a)^2} + \frac{1}{(2b)^2}$ then $\frac{2}{r^2} = \frac{1}{a^2} + \frac{1}{b^2} = \frac{b^2 + a^2}{a^2\ b^2}$

$$\text{Hence} \quad \frac{2\ (a^2\ \sin^2 + b^2\ \cos^2\theta)}{a^2\ b^2} = \frac{a^2 + b^2}{a^2\ b^2} (\sin^2\theta + \cos^2\theta)$$

$$\Rightarrow \quad 2a^2 \sin^2\theta + 2b^2 \cos^2\theta = (a^2 + b^2) \sin^2\theta + (a^2 + b^2) \cos^2\theta$$

$$\Rightarrow \quad \sin^2\theta\ (a^2 - b^2) = \cos^2\theta\ (a^2 - b^2)$$

$$\Rightarrow \quad \tan^2\theta = 1 = \text{or } \tan\theta = 1 \qquad \therefore \theta = 45^\circ$$ **Ans.**

Example 77:

The tangent at any point P of a circle meets the tangent at a fixed point. A in T, and T is joined to B, the other end of the diameter through A; prove *that the locus of the intersection of AP and BT is an ellipse whose eccentricity is $1/\sqrt{2}$.*

Solution:

Suppose the circle is $x^2 + y^2 = a^2$. Its radius is a and centre (0, 0). Let (a cos α, and sin α) be any point on the circle say P and OA as x-axis. So A is (a, 0) and therefore B will be (–a, 0).

Equation of tangent at B will be $x \cos\alpha + y \sin\alpha = a$ ...(1)

and tangent at A is $x = a$ ...(2)

Solving (1) and (2), we have

$$y = \frac{a(1-\cos\alpha)}{\sin\alpha} = \frac{2a\sin^2\alpha/2}{2\sin\alpha/2\ \cos\alpha/2} = a\tan\alpha/2. \quad ...(3)$$

So the co-ordinates of T are (a, a tan α/2).

Now equation to AP is

$$y - 0 = \frac{a\sin\alpha - 0}{\sin\alpha}(x-a) = -(x-a)\cot\frac{\alpha}{2}$$

$$\Rightarrow \quad \cot\frac{\alpha}{2} = -\frac{y}{x-a} \quad ...(4)$$

and equation of B is

$$y - 0 = \frac{a\tan\alpha/2 - 0}{a-(-a)}(x+a) = \frac{1}{2}(x+a)\tan\frac{\alpha}{2}$$

$$\Rightarrow \quad \tan\frac{\alpha}{2} = \frac{2y}{x+a} \quad ...(5)$$

To eliminate the variable α, we multiply (4) and (5), so

$$\frac{-2y^2}{x^2-a^2}\ 1 \text{ or } 2y^2 + x^2 = a^2 \text{ or } \frac{x^2}{a^2} + \frac{y^2}{a^2 - a^2/2} = 1,$$

The ellipse

which is the locus of the point of intersection of AP and BT and in clearly an ellipse.

$$\text{Again } e = \left(\frac{A^2 - B^2}{A^2}\right) = \sqrt{\left(\frac{a^2 - a^2/2}{a^2}\right)} = \sqrt{\left(1 - \frac{1}{2}\right)}\ \frac{1}{\sqrt{2}} \quad \textbf{Proved.}$$

Example 78:

Prove that the sum of the squares of the perpendiculars on any tangent from two points on the minor axis, each distant $\sqrt{(a^2 - b^2)}$ from the centre, is $2a^2$.

Solution:

Suppose the equation to the ellipse is $x^2/a^2 + y^2/b^2 = 1$. ...(1)

As the distance of the points on the minor axis are at a distance $\sqrt{(a^2 - b^2)}$ from the centre (0, 0), so the co-ordinates of the points (say P and Q) will be $\{0, \sqrt{(a^2 - b^2)}\}$ and $\{0, -\sqrt{(a^2 - b^2)}\}$.

Let $y = mx + \sqrt{(a^2m^2 + b^2)}$ be any tangent on (1). ...(2)

Sum of the squares of the perpendiculars from the points P and Q on the tangent (2) is

$$= \left\{\frac{\sqrt{(a^2m^2 + b^2)} - \sqrt{(a^2 - b^2)}}{\sqrt{(1+m^2)}}\right\}^2 + \left\{\frac{\sqrt{(a^2m^2 + b^2)} + \sqrt{(a^2 - b^2)}}{\sqrt{(1+m^2)}}\right\}^2$$

$$= \frac{2(a^2m^2 + b^2 + a^2 - b^2)}{1+m^2} = \frac{2a^2(1+m^2)}{1+m^2}$$ **Proved.**

Example 79:

Find the equations to the tangents at the ends of the latus recta of the ellipse $\frac{x^2}{a^2} + \frac{y^2}{b^2} = 1$, *and show that they pass through the intersections of the axis and the directrixes.*

Solution:

Let the equation to the ellipse be $\frac{x^2}{a^2} + \frac{y^2}{b^2} = 1$

The co-ordinates of the end point L' of latera recta LSL' are $\left(-ae, \frac{b^2}{a}\right)$ and $\left(-ae, \frac{-b^2}{a}\right)$ respectively.

Equation of tangent at point L is $\frac{-aex}{a^2} + \frac{b^2}{ab^2}y = 1$

$\Rightarrow$ $y = ex + a$ ...(1)

Equation of tangent at point L' is $\frac{-aex}{a^2} - \frac{b^2}{ab^2}y = 1$

$\Rightarrow$ $y + ex + a = 0$ ...(2)

On solving (1) and (2), the co-ordinates of point of intersection are $\left(\frac{-a}{e}, 0\right)$. Again the co-ordinates of the point of intersection of x-axis, i.e.,

$y = 0$ and directrix $x = -\frac{a}{e}$ are $\left(\frac{-a}{e}, 0\right)$

Similarly we can prove for other latus rectum.

Example 80:

Prove that the sum of the eccentric angles of the extremities of a chord, which is drawn in a given direction, is constant, and equal to twice the eccentric angle of the point at which the tangent is parallel to the given direction.

Solution:

Taking the co-ordinates of the extremities of any chord of the ellipse $x^2/a^2 + y^2/b^2 = 1$, as $(a\cos\phi_1, b\sin\phi_1)$ and $(a\cos\phi_2, b\sin\phi_2)$ respectively, where ϕ_1 and ϕ_2 are the eccentric angles of the points, the slope of the chord.

$$= \frac{b\sin\phi_1 - \sin\phi_2}{a\cos\phi_1 - \cos\phi_2} \quad \frac{2b\cos\frac{\phi_1+\phi_2}{2} - \sin\frac{\phi_1-\phi_2}{2}}{2a\sin\frac{\phi_1+\phi_2}{2} - \sin\frac{\phi_2-\phi_1}{2}} = -\frac{b}{a}\cot\frac{\phi_1+\phi_2}{2}$$

$\{\because \sin(\phi_2 - \phi_1) = -\sin(\phi_1 - \phi_2)\}$

As the direction of the chord is given to be constant, so m is constant,

or $\frac{-b}{a}\cot\frac{\phi_1+\phi_2}{2} = \text{constant}.$

or $\phi_1 + \phi_2$ is constant as a and b are constant.

Suppose the tangent $(x/a)\cos\alpha + (y/b)\sin\alpha = a$, at the point $(a\cos\alpha, b\sin\alpha)$ be parallel to the chord: then slope of this tangent must be equal to that of the chord.

$$\therefore \quad \frac{-b}{a}\cot\alpha = \frac{-b}{a}\cot\frac{\phi_1+\phi_2}{2}$$

Hence $\phi_1 + \phi_2 = 2\alpha$. **Proved.**

Example 81:

Find the locus of the point of intersection of the two straight lines

$$\frac{tx}{a} - \frac{y}{b} + t = 0 \text{ and } \frac{x}{a} + \frac{ty}{b} - 1 = 0.$$

Prove also that they meet at the point whose eccentric angle is $2\tan^{-1} t$.

Solution:

Equations of the lines are given as

$$\frac{tx}{a} - \frac{y}{b} + t = 0 \qquad ...(1)$$

and $$\frac{x}{a} + \frac{ty}{b} - 1 = 0 \qquad ...(2)$$

To find the locus of the point of intersection, we have to eliminate the variable 't' from (1) and (2). So, by (2),

$$\frac{y}{b} = \frac{t}{t}\left(1 - \frac{x}{a}\right) \text{ and by (1) } \frac{y}{b}\left(\frac{x}{a} + 1\right)$$

Multiplying, we get $\left(\frac{y}{b}\right)^2 = \left(1 - \frac{x}{a}\right)\left(1 + \frac{x}{a}\right)$ or $1 - \frac{x^2}{a^2} = \frac{y^2}{b^2}$

$\frac{x^2}{a^2} + \frac{y^2}{b^2} = 1$. This is the equation of an ellipse.

Again solving (1) and (2), we get $\frac{a(1-t^2)}{1+t^2}$. Let the abscissa of the point of intersection be a cos ϕ; then

$$x = a\cos\phi = \frac{a(1-t^2)}{1+t^2} \text{ or } \cos\phi = \frac{1-t^2}{1-t^2}$$

$$\Rightarrow \quad \frac{1-\cos\phi}{1+\cos\phi} + \frac{(1+t^2)}{(1-t^2)} - \frac{(1-t^2)}{(1+t^2)} = \frac{2t^2}{2} = t^2$$

$$\Rightarrow \quad \frac{2\sin^2\phi/2}{2\cos^2\phi/2} t^2 \text{ or } \tan^2\phi/2 = t^2$$

$$\Rightarrow \quad \tan\phi/2 = t.$$

$\Rightarrow$ Hence, $\phi = 2\tan^{-1} t$. **Proved.**

Example 82:

Find the equations to the tangents to the ellipse $4x^2 + 3y^2 = 5$ which are parallel to the straight line $y = 3x + 7$.

Find also the co-ordinates of the points of contact of the tangents which are inclined at 60° to the axis of x.

Solution:

(a) The equation to the ellipse is $4x^2 + 3y^2 = 5$

or $\dfrac{x^2}{5/4} + \dfrac{y^2}{5/3} = 1.$ ...(1)

Hence $a^2 = \dfrac{5}{4}$ and $b^2 = \dfrac{5}{3}$ So $b > a$

Equation of any line parallel to $y = 3x + 7$ is $y = 3x + c$. ...(2)

If (2) is a tangent to (1), then

$$c = \pm\sqrt{(a^2m^2 + b^2)} = \pm\sqrt{\left(\frac{5}{4}9 + \frac{5}{3}\right)} = \pm\left(\frac{155}{12}\right) = \pm\frac{1}{2}\sqrt{\frac{155}{3}}.$$

Substituting in (2), the required tangents are

$$y = 3x \pm \frac{1}{2}\sqrt{\frac{155}{3}}.$$

Ans.

(b) As $y = mx \pm \sqrt{(a^2m^2 + b^2)}$...(1) is always tangent on the given ellipse $x^2/a^2 + y^2/b^2 = 1$.

Here $m = \tan 60° = \sqrt{3}$ (given), putting in (1), the equation of tangent on ellipse is

$$y = \sqrt{3x} \pm \sqrt{\left(3 \cdot \frac{5}{4} + \frac{5}{3}\right)} = \sqrt{3x} \pm \frac{1}{2}\sqrt{\frac{65}{3}}.$$

Let (x_1, y_1) be the co-ordinates of the point of contact

Then, tangent at (x_1, y_1) w.r.t. (1) $4xx_1 + 3yy_1 = 5$. This line must be identical with (2).

$$\text{So} \quad \frac{-4x_1}{\sqrt{3}} = \frac{3y_1}{1} = \pm\frac{5}{\frac{1}{2}\sqrt{\left(\frac{65}{3}\right)}} = \pm\frac{2 \times \sqrt{3} \times \sqrt{65}}{13}$$

Hence $x_1 = \pm\dfrac{2 \times \sqrt{3} \times \sqrt{65}}{13 \times 4} = \pm\dfrac{3}{26}\sqrt{65}$ and $y_1 = \pm\dfrac{2}{39}\sqrt{195}$ or the co-ordinates of the point of contact are

$$\left(\pm\frac{3}{26}\sqrt{65}, \pm\frac{2}{39}\sqrt{195}\right)$$

Ans.

Example 83:

Find the equations to the normals at the ends of the latera recta, and prove that each passes through an end of the minor axis if $e^4 + e^2 = 1$.

Solution:

Suppose equation to the ellipse be $x^2/a^2 + y^2/b^2 = 1$ and let the end point of the latus rectum be $(ae, b^2/a)$.

The normal at this point will be

$$\frac{x - ae}{ae / a^2} = \frac{y - b^2 / e}{b^2 / a^2 / b^2} \text{ or } \frac{x - ae}{e} = y - \frac{b^2}{a}$$

If this normal passes through (0, –b) then, we have $\frac{-ae}{e} = -b\ \frac{b^2}{a}$

$\Rightarrow \quad a = b + a(1 - e^2)$ or $b - ae^2 = 0$

$\Rightarrow \quad \frac{b}{a} = e^2$ or $\frac{b^2}{a^2} = e^4$

$\Rightarrow \quad (1 - e^2) = e^4$ or $e^4 + e^2 = 1$. **Proved.**

Example 84:

In an ellipse, referred to its centre, the length of the sub-tangent corresponding to the point (3, 12/5) is 16/3 ; prove that the eccentricity is 4/5.

Solution:

Suppose the tangent at point (3, 12/5) (say P) meets the axis of the ellipse

$$\frac{x^2}{a^2} + \frac{y^2}{b^2} = 1 \text{ in T.}$$

Now, the tangent at P will be

$$\frac{3x}{a^2} + \frac{12y}{5b^2} = 1.$$

Solving it with x-axis, i.e.,

y = 0, we get $x = \frac{a^2}{3}$.

Hence $CT = \frac{a^2}{3}$. If PN is the perpendicular from P on x-axis subnormal NT = CT – CN and is clearly equal to the abscissa of P.

So $\quad \frac{a^2}{3} - 3 = \frac{16}{3}$ (given)

or $\quad a^2 = 25$

As the point $\left(3, \frac{12}{5}\right)$ lies on the ellipse, so

$$\frac{9}{25} + \frac{144}{25b^2} = 1 \text{ (as } a^2 = 25)$$

or $\quad b^2 = 9$

Now $\quad e = \sqrt{\left(\frac{a^1 - b^2}{a^2}\right)} = \sqrt{\left(\frac{25-9}{25}\right)} = \frac{4}{5}$ **Ans.**

Example 85:

Find the equation to the tangent and normal at the ends of the latera recta of the ellipse $9x^2 + 16y^2 = 144$.

Solution:

The equation of the ellipse is $9x^2 + 16y^2 = 144$

or $\quad \frac{x^2}{16} + \frac{y^2}{9} = 1$

Hence $\quad a^2 = 16$ and $b^2 = 9$

and $\quad e = \sqrt{\left(1 - \frac{b^2}{a^2}\right)} = \sqrt{\left(1 - \frac{9}{16}\right)} = \frac{\sqrt{7}}{4}$

As the co-ordinates of the ends of the latera recta are $(\pm ae, \pm b^2/a)$ from (1), the points are $(\pm\sqrt{7}, \pm 9/4)$.

The tangents at $(\pm 7, \pm 9/4)$ are given by

$$9.x\,(\pm\sqrt{7}) + (16 \pm 9/4y) = 144$$

or $\quad \pm x\sqrt{7} \pm 4y = 16.$

The slope of the tangents is $\pm\sqrt{7}/4$. Hence the slope of the normal is $\pm 4\sqrt{7}$.

So equations of the normals will be

$$(y \pm 9/4) = \pm (4/\sqrt{7})\,(x \pm\sqrt{7})$$

or $\quad 4x \pm \sqrt{7}y = 7/4\sqrt{7}.$ **Ans.**

Example 86:

Prove that the straight line $y = x + \sqrt{7/12}$ touches the ellipse $3x^2 + 4y^2 = 1$.

Solution:

The equation of the ellipse is given as

$$3x^2 + 4y^2 = 1 \quad ...(1)$$

and the equation to the line is given as

$$y = x + \sqrt{\{7/12\}} \quad ...(2)$$

Putting the value of y from (2) in (1), we have

$$3x^2 + 4 (x + \sqrt{\{7/12\}}^2 = 1$$

or $3x^2 + 4 (x^2 + 2x\sqrt{7/12} + 7/12) = 1$

or $7x^2 + 4x\sqrt{\frac{7}{3}} + \frac{4}{3} = 0$

or $(\sqrt{7}x + 2/\sqrt{3})^2 = 0 \quad ...(3)$

As (3) is a perfect square, so the roots of x are coincident; of the line (2) touches (1). **Proved.**

Example 87:

Find the points on the ellipse such than the tangent on either of them makes equal angles with the axes. Prove also that the length of the perpendicular from the centre on either of these tangents is

$$\sqrt{[a^2 + b^2/2]}.$$

Solution:

If the co-ordinates of the required point on the ellipse

$$\frac{x^2}{a^2} + \frac{y^2}{b^2} = 1$$

be (a cos ϕ, b sin ϕ), then the tangent at this point is

$$\frac{x}{a}\cos\phi + \frac{y}{b}\sin\phi = 1 \quad ...(1)$$

Slope of (1) $= \frac{-\cos\phi}{a} \times \frac{b}{\sin\phi} = -\frac{b}{a}\cot\phi = \pm\tan 45^\circ = \pm 1.$

(by hypothesis)

(as the tangent makes equal angles with the axes)

or $\cot\phi = \pm\frac{b}{a}.$

Hence $\cos\phi = \pm\frac{a}{\sqrt{(a^2 + b^2)}}$

and $$\sin\phi = \pm \frac{b}{\sqrt{(a^2+b^2)}}$$

∴ The co-ordinates of the required points are

$$\left\{\pm \frac{a^2}{\sqrt{(a^2+b^2)}}, \pm \frac{b^2}{\sqrt{(a^2+b^2)}}\right\}$$

Again the length of perpendicular from (0, 0) on (1)

$$= \frac{ab}{b^2\cos\phi + a^2\sin\phi}, = \frac{ab}{\sqrt{\left[\frac{b^2a^2}{a^2+b^2} + \frac{a^2b^2}{a^2+b^2}\right]}} = \sqrt{\frac{(a^2+b^2)}{2}}$$

Example 88:

If P be a point on the ellipse, whose ordinate is y', prove that the angle between the tangent at P and the focal distance of P is

$$\tan^{-1}\frac{b^2}{aey'}$$

Solution:

If (x', y') be the co-ordinates of the given point P on the ellipse

$$\frac{x^2}{a^2} + \frac{y^2}{b^2} = 1.$$

The tangent at (x', y') will be $\frac{xx'}{a^2} + \frac{yy'}{b^2} = 1$.

The slope of PT $\frac{b^2}{a^2}\cdot\frac{x'}{y'} = m_1$ (say).

The slope of PS $\frac{y'}{x'+ae} = m_2$ (say).

If angle between the focal distance SP and tangent PT is θ, then

$$\tan\theta = \frac{m_2 - m_1}{1 + m_1m_2} = \frac{\frac{y'}{x'+ae} + \frac{b^2}{a^2}\cdot\frac{x'}{y'}}{1 - \frac{b^2}{a^2}\cdot\frac{x'}{x'+ae}}$$

$$= \frac{(a^2y'^2 + b^2x'^2 + aeb^2x')}{(a^2x'^2 + a^3e - b')y'} = \frac{(a^2b^2 + aeb^2x')}{(a^2e^2x' + a^3e)y'} \qquad (\because a^2 - b^2 = a^2e^2)$$

and as (x', y') lies on $b^2x^2 + a^2y^2 = a^2b^2$,

we have $$b^2x'^2 + y'^2a^2 = a^2b^2$$

So $\tan\theta = \frac{ab^2(a = ex')}{a^2e(a+ex')y'} = \frac{b^2}{aey'}$

$\therefore \theta = \tan^{-1}\frac{b^2}{aey'}$ **Ans.**

Example 89:

A circle radius r, is concentric with the ellipse; prove that the common tangent is inclined to the major axis at an angle.

$$\tan^{-1}\sqrt{\frac{r^2 - b^2}{a^2 - r^2}}$$ *and find its length.*

Solution:

(a) $\frac{x^2}{a^2} + \frac{y^2}{b^2} = 1$, be the equation to the ellipse, then equation to the circle concentric to it having radius r will be

$$x^2 + y^2 = r^2.$$

As the line $y = mx + \sqrt{(a^2m^2 + b^2)}$ ...(1)

is always tangent on the ellipse, if this is also a tangent on the circle, then length of perpendicular from the centre (0, 0) on the line (1) must be equal to radius of the circle, i.e., r.

Hence $\sqrt{\frac{(a^2 m^2 + b^2)}{(1 + m^2)}} = r$

or $a^2 m^2 + b^2 + b^2 = r^2 (1 + m^2)$

or $m = \frac{r^2 - b^2}{a^2 - r^2}$ or $m = \sqrt{\frac{r^2 - b^2}{a^2 - r^2}}$

Hence, the common tangent to the two curves is inclined at an angle of

$$\tan^{-1}\sqrt{\frac{r^2 - b^2}{a^2 - r^2}}$$ to the x – axis. **Proved.**

(b) Let P and Q be the points of contact of the common tangent with the ellipse and circle respectively and O be the common centre of the two, then $PQ = \sqrt{(OP^2 - OQ^2)}$; $[\because \angle CQP = 90°]$

The co-ordinates of P are $\left(\frac{-a^2 m}{\sqrt{(a^2 m^2 + b^2)}}, \frac{b^2}{\sqrt{(a^2 m^2 + b^2)}}\right)$

So $OP^2 \dfrac{a^2 m^2 + b^2}{a^2 m^2 + b^2} = a^2 + \dfrac{b^2 \left(b^2 - a^2\right)}{a^2 m^2 + b^2}$ (by division)

Again $a^2 m^2 + b^2 = a^2 \dfrac{r^2 - b^2}{a^2 - r^2} + b^2 = \dfrac{a^2 r^2 - b^2 r^2}{a^2 - r^2}$

$$\left[\therefore \quad m^2 = \frac{r^2 - b^2}{a^2 - r^2}\right]$$

Hence $OP^2 = a^2 + \dfrac{b^2 (b^2 - a^2)(a^2 - r^2)}{r^2 \left(a^2 - b^2\right)} = a^2 - \dfrac{b^2 (a^2 - r^2)}{r^2} = r$

as $OQ = r$, so $PQ^2 = OP^2 - OQ^2$

$$= a^2 - \frac{b^2 (a^2 - r^2)}{r^2} - r^2$$

or $PQ^2 = \dfrac{(a^2 - r^2)(r^2 - b^2)}{2}$ or $PQ = \dfrac{\sqrt{(a^2 - r^2)(r^2 - b^2)}}{r}$ **Ans.**

2

HYPERBOLA

Definition : *The hyperbola is conic section in which the eccentricity e is greater than 1. "A hyperbola is the locus of a point P which moves in such a way that its distance from a fixed point (called the focus) bears a constant ratio greater than one (called the eccentricity) to its perpendicular distance from a fixed straight line (called the directrix).*

TO FIND THE EQUATION OF A HYPERBOLA

To find the equation of a hyperbola in standard form.

Let ZM be the directrix and S the focus of the hyperbola.

Let SZ be drawn perpendicular to the directrix.

Since $e > 1$, divide SZ internally and externally in the ratio $e : 1$. Let A and A' be the points of division. By definition the points A and A' lie on the hyperbola. Suppose $AA' = 2a$ and let C be the middle point of AA' so that $CA = CA' = a$.

Now we have

$$SA = e.\ AZ, \qquad ...(1)$$

and $$SA' = e.AZ'. \qquad ...(2)$$

Relation (1) and (2) may respectively be written as

and $$(CS - CA) = e.AZ, \qquad ...(3)$$

Adding (3) and (4), we have ...(4)

$$2CS = e\ (AZ + ZA') \qquad [\because\ CA = CA' = a]$$

$$\Rightarrow \qquad = e.AA' = e.2a$$

$$\Rightarrow \quad 2CS = 2ea$$

$$\Rightarrow \quad CS = ae. \qquad ...(5)$$

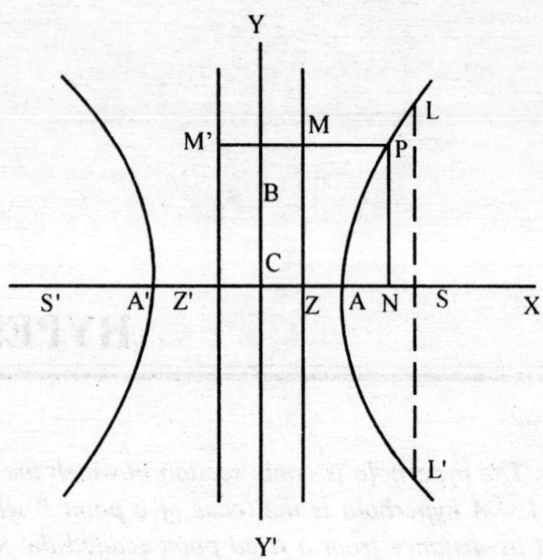

Subtracting (3) from (4)

$$CA' + CA = e\ (ZA' - ZA)$$

$$\Rightarrow \quad 2CA = e\ [CA' + CZ) - (CA - CZ)]$$

$$\Rightarrow \quad 2a = e\ [CA' + CZ) - (CA - CZ)]$$

$$\Rightarrow \quad CZ = a/e.$$

Now take C as origin, CS as the x-axis, and a perpendicular line CY as the y-axis. Then the co-ordinates of the focus S are (ae, 0) and the equation of the directrix ZM is given by x = a/e.

Consider a point P(x, y) on the hyperbola. Draw PM perpendicular too the directrix we know that

$$PS = e.\ PM \text{ or } PS^2 = e^2.\ PM^2$$

$$\Rightarrow \quad (x - ae)^2 + (y - 0)^2 - e^2\ [x - (a/e)]^2$$

$$\Rightarrow \quad (x - ae)^2 + y^2 = e^2\ [x - (a/e)]^2$$

$$\Rightarrow \quad x^2 - ae)^2 + y^2 = e^2\ [x - (a/e)]^2$$

$$\Rightarrow \quad x^2 + a^2e^2 - 2xae + y^2 = e^2\ x^2 - 2aex + a^2$$

$$\Rightarrow \quad x^2\ (e^2 - 1) - y^2 = a^2\ (e^2 - 1)$$

$$\Rightarrow \quad \frac{x^2}{a^2} - \frac{y^2}{a^2\left(e^2-1\right)} = 1. \qquad ...(7)$$

Since in the case of hyperbola e > 1. The quantity $a^2\ (e^2 - 1)$ is a positive. Let it be b^2, so that the equation (2) becomes

$$(x^2/a^2) - (y^2/b^2) = 1,$$

which is the equation of a hyperbola in standard form.

We have supposed $b^2 = a^2 (e^2 - 1)$.

Hence the eccentricity e is given by

$$e^2 = 1 + (b^2/a^2).$$

There exists a second focus and a second directrix to the curve.

On SC produce take a point S' on it such that CS = CS' = ae.

Also take another point Z' on it such that CZ = CZ' = a/e. Draw Z'M' perpendicular to S' S and produce PM to meet Z' M' in M' so that PM' is perpendicular to Z' M'.

The equation of the hyperbola is given by

$$\frac{x^2}{a^2} - \frac{y^2}{a^2(e^2-1)} = 1$$

$\Rightarrow$ $x^2(e^2 - 1) - y^2 = a^2 (e^2 - 1)$, or $x^2e^2 - x^2 - y^2 = a^2e^2 - a^2$

$\Rightarrow$ $x^2 + a^2e^2 + y^2 = x^2e^2 + a^2$.

Adding 2aex to both sides, we obtain

$$(x^2 + 2aex + a^2e^2) + y^2 = x^2e^2 + 2aex + a^2$$

$\Rightarrow$ $(x + ae)^2 + y^2 = (a + xe)^2$

$\Rightarrow$ $(x + ae)^2 + y^2 = e^2(a/e + x)^2$

$\Rightarrow$ $S'P^2 = e^2 PM'^2$, or $S'P = e. PM'$.

Thus we observe that any point P on the hyperbola is such that its distance from the point S' (on axis) is e times its perpendicular distance from the straight line Z' M'.

A GEOMETRICAL PROPERTY OF THE HYPERBOLA

The equation of the hyperbola is given as

$$x^2/a^2 - y^2/b^2 = 1 \qquad ...(1)$$

$\therefore$ $y^2/b^2 = x^2/a^2 - 1 = (x^2 - a^2)/a^2 = \{(x + a)(x - a)\}/a^2$. ...(2)}

From the figure 1 we have

PN = y, x + a = CN + CA' = A' N, x – a = CN – CA = AN.

Putting these values in (2), we obtain

$$\frac{PN^2}{b^2} = \frac{A'N.AN}{a^2} \text{ or } \frac{PN^2}{AN.NA} = \frac{b^2}{a^2} \qquad ...(3)$$

TRANSVERSE AND CONJUGATE AXES

The points A and A' are called the vertices of the hyperbola. The line AA' is called its *transverse axis*, while the line BB' is called the conjugate axis, and such that B' C = CB = b. The lengths of transverse and conjugate axes are 2a and 2b respectively.

(i) *Centre.* The point C i.e., the middle point of the transverse axis AA', is called the centre of the hyperbola because it bisects very chord of the hyperbola passing through it. The chords of the hyperbola passing through the centre C are called the diameters.

(ii) *Latus rectum.* The chord LL' perpendicular to the transverse axis and passing through the focus S is called the latus rectum its equation is $x = ae$.

Let $2l$ be the length of the latus rectum; then the point (ae, l) lies on the hyperbola $x^2/a^2 - y^2/b^2 = 1$.

$$\therefore \quad e^2 - (l^2/b^2) = 1 \Rightarrow l^2 = b^2 (e^2 - 1)$$

$$\Rightarrow \quad l^2 = b^2 . (b^2/a^2) \qquad [\because \; b^2 = a^2 (e^2 - 1)]$$

$$\Rightarrow \quad l = b^2/a = \textit{semi-latus rectum.}$$

Length of the latus rectum $2b^2/a$.

The co-ordinates of the end points L and L' of the latus rectum LL' are $(ae, b^2/a)$ and $(ae, -b^2/a)$ respectively.

RELATION BETWEEN FOCAL DISTANCES OF A POINT

To show that the difference of the focal distances of a point on the hyperbola is constant.

Let $P(x_1, y_1)$ be a point on the hyperbola. Let S and S' be the two foci. Let ZM be the directrix corresponding to the focus S and Z' M' the directrix corresponding to the focus S' PM is perpendicular to ZM and PM' is perpendicular to Z' M'.

We have PS = e. PM = e. ZN = e (CN – CZ) = $e (x_1 - a/e) = \mathbf{ex_1 - a}$, and PS' = e. PM' = e (CN + CZ') = $e (x_1 - a/e) = \mathbf{ex_1 + a}$.

Thus PS' – PS = 2a,

which is constant and is equal to the length of the transverse axis.

TRACING THE HYPERBOLA $x^2/a^2 - y^2/b^2 = 1$

(i) The equation contains only even powers of both x and y and hence the curve is symmetrical about both the axes. Thus it is sufficient to know the shape of the curve in the first quadrant, say.

(ii) Solving the equation for y, we have

$$y = \pm b\sqrt{\{(x^2/a^2) - 1\}}.$$

If the value of x lies between $0 \le x < a$, the corresponding value of y is imaginary. Thus no part of the curve exists between the lines x = 0 and x = a. At x = a, y = 0. The curve thus meets the a-axis at the point (a, 0). At x = 0, y is imaginary and so the curve does not cut the y-axis.

As x in creases, y also increases and as $x \to \infty$, y also $\to \infty$, considering only the positive values of y.

By plotting a few more points, we see that the shape of the curve is as shown in the figure of 1.

We observe that the curve consists of two infinite branches detached from each other.

EQUATION IN POLARS

Putting $x = r\cos\theta$ and $y = r\sin\theta$, the equation $x^2/a^2 - y^2/b^2 = 1$ of the hyperbola then we have

$$\frac{1}{r^2} = \frac{\cos^2\theta}{a^2} - \frac{\sin^2\theta}{b^2}. \qquad ...(1)$$

Now we shall show *that the sum of the squares of the reciprocals of two perpendicular central radii vectors of a hyperbola is constant.*

Putting $\theta = \theta + \frac{1}{2}\pi$ and $r = r_1$ in (1), we get

$$\frac{1}{r_1^2} \frac{\cos^2\left(\theta+\frac{1}{2}\pi\right)}{a^2} = \frac{\sin^2\left(\theta+\frac{1}{2}\pi\right)}{b^2}$$

or
$$\frac{1}{r_1^2} = \frac{\sin^2\theta}{a^2} - \frac{\cos^2\theta}{a^2} \qquad ...(2)$$

Adding (1) and (2), we obtain

$$1/r^2 + 1/r^2 = 1/r^2 = \text{constant}.$$

VARIOUS RESULTS FOR HYPERBOLA

Most of the results obtained for the ellipse $x^2/a^2 + y^2/b^2 = 1$ hold good for the hyperbola $x^2/\bar{a}^2 - y^2/b^2 = 1$ is replaced by $-b$. We give below the list of results applicable in the case of the hyperbola is given by

$$x^2/a^2 - y^2/b^2 = 1$$

(i) The equation of the *tangent at any point* (x_1, y_1) on the curve is given by $$xx_1/a^2 - yy_1/b^2 = 1.$$

(ii) The equation of the *normal at any point* (x_1, y_1) on the curve is given as $$\frac{x - x_1}{x_1 / a^2} = \frac{y - y_1}{y_1 / b^2}$$

(iii) The condition that the line $y = mx + c$ is a *tangent to the hyperbola* is $c^2 = a^2m^2 - b^2$. The point of contact in this case is
$$(-a^2m/c, \ - b^2/c).$$
Any tangent to the hyperbola is $y = mx + \sqrt{(a^2m^2 - b^2)}$.

This is the equation of a tangent to the hyperbola in terms of its gradient.

The line $x \cos \alpha + y \sin \alpha = p$ *will touch* the hyperbola
$$a^2 \cos^2 \alpha - b^2 \sin^2 \alpha = p^2.$$

(iv) *The point (x_1, y_1) is outside, on or inside* the hyperbola according as $x_1^2/a^2 - y_1^2/b^2 - 1 <, =$ or > 0.

(v) The equation of the *chord of contact* of tangent drawn from the point (x_1, y_1) is $xx_1/a^2 - yy_1/b^2 = 1$.

(vi) The equation of the *polar* of any point (x_1, y_1) with respect to the hyperbola is $xx_1/a^2 - yy_1/b^2 = 1$.

(vii) The co-ordinates of the *pole* of the line $lx + my + n = 0$ w.r.t. the hyperbola are $(-a^2/n, b^2m/n)$.

DIRECTOR CIRCLE

The locus of the point of intersection of two perpendicular tangents to a hyperbola is a circle which is called the director circle.

Let the equation of the hyperbola be $x^2/a^2 - y^2/b^2 = 1$.

The equation of any tangent to the hyperbola in terms of its gradient m is given as
$$y = mx + \sqrt{(a^2m^2 - b^2)}$$
i.e., $$y - mx = \sqrt{(a^2m^2 - b^2)} \qquad ...(1)$$
and the perpendicular tangent (replacing m by $-1/m$) is
$$my + x = \sqrt{(a^2 - b^2m^2)} \qquad ...(2)$$
Squaring and adding (1) and (2), we obtain
$$y^2 (1 + m^2) + x^2 (m^2 + 1) = a^2 (1 + m^2) - b^2 (1 + m^2)$$

or $x^2 + y^2 = a^2 - b^2$.

This being free from m is the required locus, called the director circle.

CONJUGATE POINT

Let $P(x_1, y_1)$ and $Q(x_2, y_1)$ be two points. If the polar of P passes through Q, then the polar of Q also passes through P. Such points are called *conjugate points*.

The condition for the points P and Q to the conjugate w.r.t. the hyperbola is given as

$$x_1x_2/a^2 - y_1y_2/b^2 = 1.$$

CONJUGATE LINES

Let $lx + my = n$ ...(1) and $l' + m'y = n'$ ...(2) be two lines. If the pole of (1) lies on (2), then the pole of (2) also lies on (1). Such lines are called conjugate lines.

The condition that the lines (1) and (2) are conjugate w.r.t. the hyperbola is $a^2ll' - b^2mm' = nn'$.

(i) *The locus of the middle points of a system of parallel chords of a hyperbola is a diameter of the hyperbola i.e., a straight line passing through the centre of the hyperbola.*

(ii) *Auxiliary circle.* The circle described on the transverse axis as diameter is said to be the auxiliary circle of the hyperbola and its equation is given by $x^2 + y^2 = a^2$.

(iii) *Pair of tangents.* The equation of the pair of tangents from the point (x_1, y_1) to the hyperbola is $SS_1 = T^2$

i.e., $(x^2/a^2 - y^2/b^2 - 1)(x_1^2/a^2 - y_1^2/b^2 - 1) = (xx_1/a^2 - yy_1/b^2-1)^2$.

(iv) The equation of the chord whose middle point is (x_1, y_2) is T = S, where

$$T \equiv xx_1/a^2 - yy_1/b^2 - 1$$

and $$S_1 \equiv x_1^2/a^2 - y_1^2/b^2 - 1.$$

Thus the equation of the chord whose middle point is (x_1, y_1) is $xx_1/a^2 - yy_1/b^2 = x_1^2/a^2 - y_1^2/b^2$.

Let the equation of a system of parallel chords be given by

$$y = mx + c$$

where m is a constant and c is a parameter *i.e.* c can take nay real value. However, for a particular chord belonging to the system (1), c is fixed.

Let the equation of a hyperbola be given as

$$x^2/a^2 - y^2/b^2 = 1. \quad ...(2)$$

Eliminating y between (1) and (2), the abscissae of the points of intersection of (1) and (2) then we have

$$x^2/a^2 - (mx + c)^2/b^2 = 1,$$

$$\Rightarrow \quad b^2x^2 - a^2(m^2x^2 + 2mcx + c^2) = a^2b^2$$

$$\Rightarrow \quad x^2 - (a^2m^2 - b^2) + 2a^2mcx + a^2(c^2 + b^2) = 0 \quad ...(3)$$

Thus if (x_1, y_1) and (x_2, x_2) are the points of intersection of (1) and (2), then x_1, x_2 are the roots of the quadratic (3) in x.

Now if (h, k) be the co-ordinates of the middle point of the chord (1), then

$$2h = x_1 + x_2 = (-2a^2mc)/(a^2m^2 - b^2). \quad ...(4)$$

But the point (h, k) lies on (1). Therefore $k - mh = c$. ...(5)

Eliminating the parameter c between (4) and (5), then we get

$$h(a^2m^2 - b^2) = -a^2m(k - mh)$$

or $\quad hb^2 = a^2mk$

or $\quad k = (b^2/a^2m)h.$

Thus the locus of the point (h, k) for varying values of c is given as

$$y = (b^2/a^2m)x, \quad ...(6)$$

which is a straight line passing through the centre (0, 0) of the hyperbola and so is a diameter of the hyperbola.

If we write the equation (6) in the form

$$y = m'x,$$

then $\quad m' = b^2/a^2m$

or $\quad mm' = b^2/a^2.$ ...(7)

Thus the diameter $y = m'x$ bisects the chords of the hyperbola which are parallel to the diameter

$$y = mx \text{ if } mm' = b^2/a^2.$$

A PROPERTY OF CONJUGATE DIAMETERS

To prove that of a pair of conjugate diameters of a hyperbola, only one meets the curve in real points.

Let the diameters $\quad y = m_1x$ ...(1)

and $\quad y = m_2x$ ...(2)

are conjugate diameters of the given hyperbola

$$x^2/a^2 - y^2/b^2 = 1. \qquad ..(3)$$

Then $\quad m_1 m_2 = b^2/a^2. \qquad ...(4)$

The diameter (1) cuts the hyperbola (3) in points whose abscissae are given by

$$\frac{x^2}{a^2} - \frac{m_1{}^2 x^2}{b^2} = 1 \text{ i.e., } x^2 = \frac{a^2b^2}{b^2 - a^2 m_1{}^2} \qquad ...(5)$$

Similarly the diameter (2) cuts (3) at points whose abscissae are given by

$$x^2 = \frac{a^2b^2}{b^2 - a^2 m_2{}^2} = \frac{a^2b^2}{b^2 - a^2.(b^4 - a^4 m_1{}^2)} \text{ substituting for } m_2 \text{ from (4)}$$

we have
$$= \frac{a^4 b_1{}^2}{a^2 m_1{}^2 - b^2}$$

Now the values of x given by (5) are real if the value of x^2 given by (5) is positive i.e., if $b^2 - a^2 m_1{}^2 > 0$. But then $a^2 m_1{}^2 - b^2 < 0$ and so the values of x given by (6) are real, then the values of x given by (5) are imaginary. Hence of a pair of conjugate diameters of a hyperbola, only one meets the curve in real points.

Example 1:

Show that the polar of any point on the ellipse $x^2/a^2 + y^2/b^2 = 1$ with respect to the hyperbola $x^2/a^2 - y^2/b^2 = 1$ will touch the ellipse at the other end of the ordinate through the point.

Solution:

Let PQ be a double ordinate of the given ellipse. If P be (x_1, y_1), then Q is $(x_1, -y_1)$. The equation of polar of $P(x_1, y_1)$ w.r.t. the given hyperbola is

$$xx_1/a^2 - yy_1/b^2 = 1. \qquad ...(1)$$

Again equation of the tangent at $Q(x_1, -y_1)$ to the given ellipse is

$$xx_1/a^2 - yy_1/b^2 = 1. \qquad ...(2)$$

Equations (1) and (2) are the same. Hence the required result follows.

Example 2:

Show that the locus of the mid points of the chords of the hyperbola

$$x^2/a^2 - y^2/b^2 = 1$$

which subtend a right angle at the centre of the curve is

$$(x^2/a^2 - y^2/b^2)^2 \, (1/a^2 - 1/b^2) = x^2/a^4 + y^2/b^4.$$

Solution:

Let C be the centre and PQ a chord of the hyperbola

$$x^2/a^2 - y^2/b^2 = 1. \quad ...(1)$$

If (x_1, y_1) is the mid point of PQ, then its equation by T = S_1 is

$$xx_1/a^2 - yy_1/b^2 = x_1^2/a^2 - y_1^2/b^2. \quad ...(2)$$

Now the point C is the origin. Therefore the combined equation of the lines CP and CQ joining the origin C to the points of intersection of (2) and (1) is obtained by making the equation (1) homogeneous with the help of (2). So it is given by

$$\frac{x^2}{a^2} - \frac{y^2}{b^2} = \left\{\frac{xx_1/a^2 - yy_1/b^2}{x_1^2/a^2 - y_1^2/b^2}\right\}^2$$

$$\Rightarrow (x^2/a^2 - y^2/b^2)(x_1^2/a^2 - y_1^2/b^2) - (xx_1/a^2 - yy_1/b^2)^2 = 0. \quad ...(3)$$

Now PQ subtends a right angle at the centre C (0, 0). Therefore the lines CP and CQ given by (3) are at right angles. Hence in the equation (3).

The coefficient of x^2 + the coefficient of y^2 = 0

$$\Rightarrow \left\{\frac{1}{a^2}\left(\frac{x_1^2}{a^2} - \frac{y_1^2}{b^2}\right)^2 - \frac{x_1^2}{a^4}\right\} + \left\{-\frac{1}{b^2}\left(\frac{x_1^2}{a^2} - \frac{y_1^2}{b^2}\right)^2 - \frac{y_1^2}{b^4}\right\} = 0$$

$$\Rightarrow (x_1^2/a^2 - y_1^2/b^2)^2 (1/a^2 - 1/b^2) = x_1^2/a^4 + y_1^2/b^4.$$

$\therefore$ the locus of (x_1, y_1) is

$$(x^2/a^2 - y^2/b^2)^2 (1/a^2 - 1/b^2) = x^2/a^4 + y_1^2/b^4.$$

Example 3:

Find the equation to hyperbola whose directrix is 2x + y = 1, focus (1, 2) and eccentricity √3.

Solution:

Consider any (x, y) on the curve.

Distance between (x, y) and the focus (1, 2)

$$= \sqrt{\{(x-1)^2 + (y-2)^2\}}.$$

Perpendicular distance of (x, y) from the directrix 2x + y – 1 = 0 is

$$= (2x + y - 1)/\sqrt{(4 + 1)}.$$

But the definition of a hyperbola, we have the distance of the point (x, y) on the curve from the focus

= × [distance of (x, y) from the directrix]

i.e., $\sqrt{\{(x-1)^2 + (y-2)^2\}} = \sqrt{3} + \{(2x + y - 1)/\sqrt{5}\}$

Squaring both sides, we get

$$x^2 - 2x + 1 + y^2 + 4 - 4y = \frac{3}{5}\{4x^2 + y^2 + 1 + 4xy - 4x - 2y\}$$

$\Rightarrow 7x^2 + 12xy - 2y^2 - 2x = 14y - 22 = 0.$

This is the required equation of the hyperbola.

Example 4:

A series of chords of the hyperbola $x^2/a^2 - y^2/b^2 = 1$ touch the circle on the line joining the foci as diameter. Show that the locus of the poles of these chords with respect to the hyperbola is the ellipse $x^2/a^4 + y^2/b^2 = 1/(a^2 + b^2)$.

Solution:

The given hyperbola is $x^2/a^2 + y^2/b^2 = 1$. ...(1)

The foci of the hyperbola (1) are S (ae, 0) and S' (–ae, 0).

The equation of the circle on SS' as diameter is

$$(x - ae)(x + ae) + (y = 0) = 0$$

i.e., $x^2 + y^2 = a^2e^2$. ...(2)

Let (x_1, y_1) be the pole of a chord of (1) w.r.t. (1). Then this chord is the polar of the point (x_1, y_1) with respect to the hyperbola (1) and so its equation is

$$xx_1/a^2 - yy_1/b^2 = 1. \quad ...(3)$$

We have to find the locus of the point (x_1, y_1) if the straight line (3) touches the circle (2).

The line (3) will touch the circle (2) if the length of the perpendicular drawn from the centre (0, 0) of the circle to the line in equal to the radius ae of the circle. Hence,

$$1/\sqrt{(x_1^2/a^4 + y_1^2/b^4)} = ae$$

$$\Rightarrow \quad \frac{x_1^2}{a^2} + \frac{y_1^2}{b^2} = \frac{1}{a^2e^2} = \frac{a^2b^2}{a^2\left(1 - b^2/a^2\right)} = \frac{1}{a^2 + b^2}$$

$\therefore$ the locus of the pole (x_1, y_1) is

$$x^2/a^4 + y^2/b^4 = 1/(a^2 = b^2), \text{ which is an ellipse.}$$

ASYMPTOTES

Definition: *An asymptote is a straight line which meets the conic in two points both of which are situated at an infinite distance, but which is itself not altogether at infinity.*

To find the asymptotes of the hyperbola

Let the equation of the hyperbola be given as

$$x^2/a^2 - y^2/b^2 = 1. \qquad \text{...(1)}$$

The tangent to the curve (1) at the point (x_1, y_1) on it is given by

$$xx_1/a^2 = yy_1/b^2 = 1 \qquad \text{...(2)}$$

Since the point (x_1, y_1) lies on (1), therefore we have

$$x_1^2/a^2 - y_1^2/b^2 = 1. \qquad \text{...(3)}$$

From (3), $y_1^2/b^2 = x_1^2/a^2 = 1$ or $y_1 = \pm b\sqrt{(x_1^2/a^2 - 1)}$.

Putting this value of y_1 in (2), we find that the equations of the tangents to (1) at the two points on it with abscissa x_1, are given by

$$\frac{xx_1}{a^2} \pm \frac{y}{b}\sqrt{\left(\frac{x_1^2}{a^2} - 1\right)} = 1$$

i.e., $$\frac{x}{a^2} \pm \frac{y}{b}\sqrt{\left(1 - \frac{a^2}{x_1^2}\right)} = \frac{a}{x_1} \qquad \text{...(4)}$$

[Dividing both sides by x_1/a]

Taking limits when x_1 tends to infinity, we see that the equations (4) tend to

$$x/a \pm y/b = 0$$

or $$y = \pm (b/a)\, x. \qquad \text{...(5)}$$

Hence $x/a + y/b = 0$ and $x/a - y/b = 0$ are the equations of the two asymptotes of (1) and the combined equation of the asymptotes is $x^2/a^2 - y^2/b^2 = 0$. *The combined equation of the asymptotes differs from the equation (1) of the hyperbola by simply a constant term.*

Remarks:

(i) The angle between the asymptotes given by (5)

$$= \tan^{-1} \frac{(b/a)+(b/a)}{1-(b/a)(b/a)} = \tan^{-1} \frac{2(b/a)}{1-(b^2/a^2)} = \tan^{-1}(b/a).$$

(ii) The equation of the pair of tangents from the centre (0, 0), to the hyperbola is $SS_1 = T^2$

$$\Rightarrow \left(\frac{x^2}{a^2} - \frac{y^2}{b^2} - 1\right)(0 - 0 - 1) = \left(\frac{0.x}{a^2} - \frac{0.y}{b^2} - 1\right)^2$$

$\Rightarrow \quad x^2/a^2 - y^2/b^2 = 0.$

Hence the pair of tangents drawn to the hyperbola from the centre are the asymptotes of the hyperbola

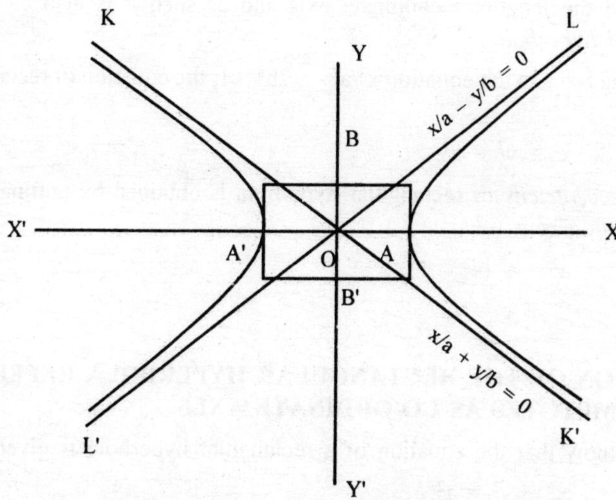

(iii) *It is easily seen that the asymptotes lie along the diagonals of the rectangle formed by drawing lines through B, B' parallel to transverse axis and through A, A' parallel to conjugate axis.*

(iv) Clearly the asymptotes (5) pass through the centre (0, 0) of the hyperbola and the axes of the hyperbola (i.e., the transverse axis and the conjugate axis) bisect the angles between them.

RECTANGULAR HYPERBOLA

Definition: *A hyperbola whose asymptotes are at right angles to each other is known as be a rectangular hyperbola.*

To we know that the equation of a rectangular hyperbola

The angle between the asymptotes of the hyperbola $x^2/a^2 - y^2/b^2 = 1$ is $2 \tan^{-1} (b/a)$. We have

If the hyperbola is rectangular, then

$$2 \tan^{-1} (b/a) = \frac{1}{2}\pi$$

i.e., $$\tan^{-1}(b/a) = \frac{1}{4}\pi$$

i.e., $\tan \frac{1}{4}\pi = b/a$ i.e., $b = a$.

Thus, we see that in a rectangular hyperbola the length of transverse axis is equal to the length of conjugate axis and as such it is also called an *equilateral hyperbola.*

Putting $b = a$ in the equation $x^2/a^2 - y^2/b^2 = 1$, the equation of rectangular hyperbola

$$x^2 = y^2 = a^2.$$

The *eccentricity* of rectangular hyperbola is obtained by putting $b = a$ in the relation $b^2 = (e^2 - 1)$

$$\therefore \quad a^2 = a^2 (e^2 - 1)$$

$$\Rightarrow \quad 2 \Rightarrow e = \sqrt{2}.$$

EQUATION OF THE RECTANGULAR HYPERBOLA REFERRED TO ASYMPTOTES AS CO-ORDINATE AXES

We know that the equation of a rectangular hyperbola is given by

$$x^2 + y^2 = a^2. \qquad ...(1)$$

Herc its transverse and conjugate axes are considered as the axes of reference and its asymptotes are at right angles. Thus each asymptote is inclined at an angle of 45° to the transverse axis i.e., the x-axis.

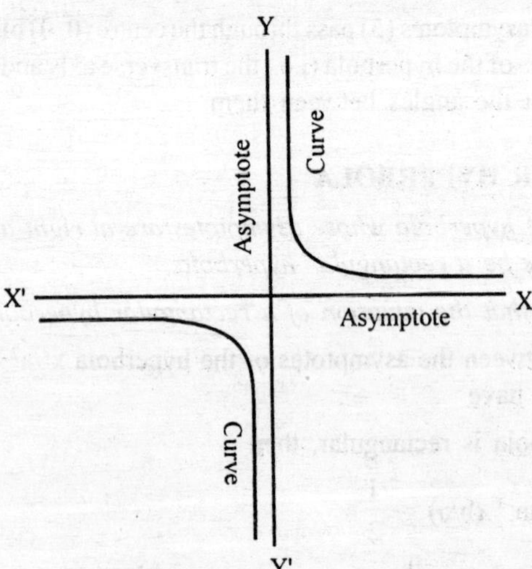

So the equation of the rectangular hyperbola with reference to its asymptotes as co-ordinate axis will be obtained by turning the co-ordinates axes through an angle of 45°, say, in the clockwise direction. Hence rotating the co-ordinate axes through an angle –45° by putting x cos (–45°) for x and y cos (–45°) + x sin (–45°) for in (1), we have

$$\{x \cos(-45^\circ) - y \sin(-45^\circ)\}^2 - \{y \cos(-45^\circ) + x \sin(-45^\circ)\}^2 = a^2$$

$$\Rightarrow \quad (x/\sqrt{2} + y/\sqrt{2})^2 - (y/\sqrt{2} - x/\sqrt{2})^2 = a^2$$

$$\Rightarrow \quad 2xy = a^2$$

$$\Rightarrow \quad xy = \frac{1}{2} a^2 \qquad ...(2)$$

Putting $\frac{1}{2} a^2 = c^2$, the equation (2) is generally written as

$$xy = c^2. \qquad ...(3)$$

EQUATION OF A GENERAL HYPERBOLA REFERRED TO ASYMPTOTES AS AXES OF CO-ORDINATES

The equation of any hyperbola referred to its transverse axis ax x-axis and its conjugate axis as y-axis is given by

$$x^2/a^2 - y^2/b^2 = 1. \qquad ...(1)$$

Let P be any point on (1). From P draw PH parallel to one asymptote CL meeting the other CK' in H and suppose CH = h and HP = k. Then the co-ordinates of P referred to the asymptotes as oblique axes are (h, k).

Now suppose θ is the half of the angle between the asymptotes, so that tan θ = b/a. Then we have

$$\therefore \quad \frac{\sin\theta}{b} = \frac{\cos\theta}{a} = \frac{1}{\sqrt{(a^2 + b^2)}}. \qquad ...(2)$$

Again draw HN perpendicular to the transverse axis, and HR parallel to the transverse axis to meet the ordinate PM produced in R.

By construction PH is parallel to CL and HR is parallel to CM, therefore ∠PHR = ∠LCM = θ.

Let the co-ordinates of P referred to CX and CY as co-ordinate axes be (x, y). Then we have

$$x = CM = CN + NM = CN + HR = CH \cos\theta + PH \cos\theta$$

$$= h \cos\theta + k \cos\theta = (h + k)\{a/\sqrt{(a^2 + b^2)}\} \qquad \text{(by 2)}$$

$$x = PM = PR + RM = OR = NH = PH \sin\theta + CH \sin\theta$$

$$= k \sin\theta - h \sin\theta = (k - h).\{b/\sqrt{(a^2 + b^2)}\} \qquad \text{(by 2)}$$

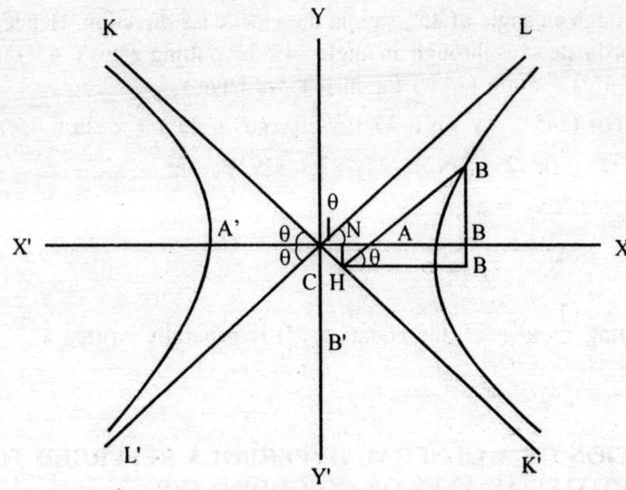

Substituting the values of x and y in (1), we have

$$\frac{(h+k)^2}{a^2+b^2} - \frac{(k-h)^2}{a^2+b^2} = 1$$

i.e., $\quad 4hk + a^2 + b^2$

$$\Rightarrow \quad hk = \frac{1}{4}(a^2 + b^2). \qquad ...(3)$$

The locus of (h, k) is given by $xy = \frac{1}{4}(a^2 + b^2)$. ...(4)

The equation (4) is the required equation of the hyperbola with asymptotes as the oblique co-ordinate axes.

Let $a^2 + b^2 = 4c^2$. Then the equation (4) may be written as

$$\mathbf{xy = c^2}. \qquad ...(5)$$

PARAMETRIC REPRESENTATION

(i) If the equation of the hyperbola is given by

$$x^2/a^2 - y^2/b^2 = 1, \qquad ...(1)$$

then $x = a \sec \theta$, $y = b \tan \theta$ satisfy the equation (1) for all values of θ because $\sec^2 \theta = 1$. Hence $(a \sec \theta, b \tan \theta)$ are taken as the parametric coordinates of any point on (1) and q is called a parameter. This point is usually called the point 'θ'.

(ii) If the equation of a hyperbola (rectangular or otherwise) is $xy = c^2$, then $x = ct$, $y = c/t$ satisfy the equation $xy = c^2$ for all values of t (except t = 0). Hence (ct, c/t) are taken as the parametric coordinates of any point on $xy = c^2$ and t is called a parameter.

EQUATIONS OF THE CHORD, TANGENT AND NORMAL FOR THE HYPERBOLA $x^2/a^2 - y^2/b^2 = 1$ IN TERMS OF PARAMETRIC CO-ORDINATES

Equation of the Chord : Let PQ be a chord of the given hyperbola. Let the parametric coordinates of P and Q be $(a \sec \theta_1, b \tan \theta_1)$ and $(a \sec \theta_2, b \tan \theta_2)$ respectively. Then the equation of the chord PQ is given by the equation

$$y - b \tan \theta_1 = \frac{b(\tan\theta_2 - \tan\theta_1)}{a(\sec\theta_2 - \sec\theta_1)} (x - a \sec \theta_1)$$

$$\Rightarrow \quad \frac{y}{b} - \tan\theta_1 = \frac{(\sin\theta_2 / \cos\theta_2) - (\sin\theta_1 / \cos\theta_1)}{(1/\cos\theta_2) - (1/\cos\theta_1)} \left(\frac{x}{a} - \sec\theta_1\right)$$

$$\Rightarrow \quad \frac{y}{b} - \tan\theta_1 = \frac{\sin\theta_2 \cos\theta_1 - \cos\theta_2 \sin\theta_1}{\cos\theta_1 - \cos\theta_2} \left(\frac{x}{a} - \sec\theta_1\right)$$

$$\Rightarrow \quad \frac{y}{b} - \frac{\sin\theta_1}{\cos\theta_1} = \frac{-\sin(\theta_1 - \theta_2)}{-(\cos\theta_2 - \cos\theta_1)} \left(\frac{x}{a} - \sec\theta_1\right)$$

$$\Rightarrow \quad \frac{y}{b} - \frac{\sin\theta_1}{\cos\theta_1} = \frac{2\sin\frac{1}{2}(\theta_1 - \theta_2)\cos\frac{1}{2}(\theta_1 - \theta_2)}{2\sin\frac{1}{2}(\theta_1 + \theta_2)\sin\frac{1}{2}(\theta_1 - \theta_2)} \left(\frac{x}{a} - \frac{1}{\cos\theta_1}\right)$$

$$\Rightarrow \quad \frac{y}{b}\sin\left(\frac{\theta_1 + \theta_2}{2}\right) - \frac{\sin\theta_1}{\cos\theta_1}\sin\left(\frac{\theta_1 + \theta_2}{2}\right)$$

$$= \frac{x}{a}\cos\left(\frac{\theta_1 - \theta_2}{2}\right) - \frac{1}{\cos\theta_1}\cos\left(\frac{\theta_1 - \theta_2}{2}\right)$$

$$\Rightarrow \quad \frac{x}{a}\cos\left(\frac{\theta_1 - \theta_2}{2}\right) - \frac{y}{b}\sin\left(\frac{\theta_1 - \theta_2}{2}\right)$$

$$\Rightarrow \quad = \frac{1}{\cos\theta_1}\left[\cos\left(\frac{\theta_1 - \theta_2}{2}\right) - \sin\theta_1 \sin\left(\frac{\theta_1 + \theta_2}{2}\right)\right]$$

$$\Rightarrow \quad = \frac{1}{\cos\theta_1}\left[\cos\left(\theta_1 \frac{\theta_1+\theta_2}{2}\right) - \sin\theta_1 \sin\left(\frac{\theta_1+\theta_2}{2}\right)\right]$$

$$\Rightarrow \quad = \frac{1}{\cos\theta_1}\left[\cos\theta_1 \cos\left(\frac{\theta_1+\theta_2}{2}\right) + \sin\theta_1 \sin\left(\frac{\theta_1+\theta_2}{2}\right)\right.$$

$$\left. - \sin\theta_1 \sin\left(\frac{\theta_1+\theta_2}{2}\right)\right]$$

$$= \cos\frac{1}{2}(\theta_1 + \theta_2)$$

Hence the equation of the chord joining the points 'θ_1' and 'θ_2' is given as follows

$$= \frac{x}{a}\cos\left(\frac{\theta_1-\theta_2}{2}\right) - \frac{y}{b}\sin\left(\frac{\theta_1+\theta_2}{2}\right) = \cos\left(\frac{\theta_1+\theta_2}{2}\right). \qquad ...(1)$$

Tangent. As $Q \to P$, the chord $PQ \to$ the tangent at P. Therefore taking limits of both sides of (1) as $\theta_2 \to \theta_1$, we get the equation of the tangent at the point $(a \sec \theta_1, b \tan \theta_1)$ as

$$(x/a) - (y/b)\sin\theta_1 = \cos\theta_2$$

$$\Rightarrow \quad \mathbf{(x/a)\sec\theta_1 - (y/b)\tan\theta_1 = 1.} \qquad ...(2)$$

Normal : The gradient of the tangent (2) = $(b \sec \theta_1)/(b \tan \theta_1)$.

$\therefore$ the gradient of the normal at the point 'θ_1' is given as

$$= -\frac{a\tan\theta_1}{b\sec\theta_1} = -\frac{a}{b}\sin\theta_1.$$

$\therefore$ the gradient of the normal at the point $(a \sec \theta_1, b \tan \theta_1)$ is given

as $\quad y - b\tan\theta_1 = -\dfrac{a\sin\theta_1}{b}(x - a\sin\theta_1)$

$$\Rightarrow \quad ax\sin\theta + by = (a^2 + b^2)\tan\theta_1.$$

Dividing both sides by $\tan \theta_1$, we get

$$ax\cos\theta_1 + by\cot\theta_1 = a^2 + b^2, \qquad ...(3)$$

which is the required equ- of normal

Deductions. For the rectangular hyperbola $x^2 - y^2 = a^2$, putting $b = a$ in the above results, we get

the chord joining the points 'θ_1' and 'θ_2' as follows

$$x\cos\frac{\theta_1-\theta_2}{2} - y\sin\left(\frac{\theta_1+\theta_2}{2}\right) = a\cos\left(\frac{\theta_1+\theta_2}{2}\right), \qquad ...(4)$$

the tangent at 'θ_1' as $x \sec\theta_1 - y\tan\theta_1 = a$, ...(5)

and the normal at 'θ_1' as $x\cos\theta_1 + y\cot\theta_1 = 2a$. ...(6)

EQUATION OF THE CHORD, TANGENT AND NORMAL FOR THE RECTANGULAR HYPERBOLA $xy = c^2$ IN TERMS OF A PARAMETER, SAY, t

The coordinates of any point (x, y) on the hyperbola $xy = c^2$ may be taken as $x = ct$, $y = c/t$, where t is a parameter. We say it as the point 't'.

Let $P(ct_1, c/t_1)$ and $Q(ct_2, c/t_2)$ be any two points on the curve

$$xy = c^2. \quad ...(7)$$

The equation of the chord PQ is given as

$$y - \frac{c}{t_1} = \frac{(c/t_1 - c/t_1)}{c(t_2 - t_1)}(x - ct_1)$$

$$\Rightarrow \quad y - \frac{c}{t_1} = \frac{c(t_1 - t_2)}{ct_1t_2(t_2 - t_1)}(x - ct_1)$$

$$\Rightarrow \quad y - \frac{c}{t_1} = -\frac{1}{t_1t_2}(c - ct_1)$$

$$\Rightarrow \quad t_1t_2\, y - ct_2 = -x + ct_1$$

$$\Rightarrow \quad x + y\,t_1t_2 = c\,(t_1 + t_2) \quad ...(8)$$

Now taking limits of both sides of equ. (8) as $t_2 \to t_1$, the *equation of the tangent* at $P(ct_1, c/t_1)$ is given by

$$\mathbf{x + yt_1^2 = 2c\ t_1.} \quad ...(9)$$

Normal at point 't_1'. Slope of the tangent (9) = $-1/t_1^2$.

$\therefore$ Slope of the normal at the point $(ct_1, c/t_1) = t_1^2$.

Hence the equation of the normal at $(ct_1, c/t_1)$ is given by

$$y - c/t_1 = t_1^2\,(x - ct_1)$$

$$\Rightarrow \quad \mathbf{xt_1^3 - yt_1 - ct_1^4 + c = 0} \quad ...(10)$$

Deductions. Multiplying the equation (9) of the tangent by c/t_1, we have

$$x\,(c/t_1) + y\,(ct_1) = 2c^2.$$

If the point $(c/t_1, c/t_1)$ is considered at the point (x_1, y_1), the we have

$$xy_1 + yx_1 = 2c^2 = 2x_1\,y \quad \text{[by (7)]}$$

$\Rightarrow \qquad x/x_1 + y/y_1 = 2.$ ...(11)

Which is the equation of the tangent at the point (x_1, y_1) on the hyperbola (7).

CONJUGATE HYPERBOLA

Definition: *The hyperbola whose transverse and conjugate axes are respectively the conjugate and transverse axes of a given hyperbola is called the conjugate hyperbola of the given hyperbola and the two hyperbolas are known as conjugate hyperbolas.*

It the equ. of given hyperbola as $\frac{x^2}{a^2} - \frac{y^2}{b^2} = 1$, ...(1)

Then equ. of its conjugate hyperbola is $\frac{-x^2}{a^2} - \frac{y^2}{b^2} = 1$

or $\qquad \frac{x^2}{a^2} - \frac{y^2}{b^2} = -1$

Clearly we can also say that the conjugate hyperbola of (2) is

(i) Any point 'θ' on (2) is taken as $(a \tan\theta, b \sec\theta)$.

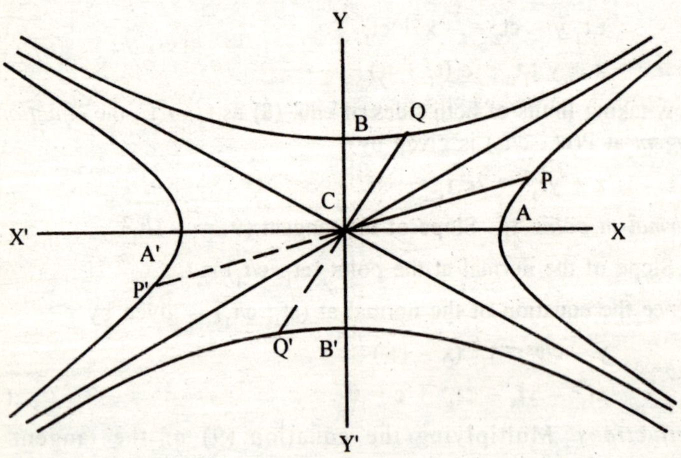

Remark:

Equation of the conjugate hyperbola of the hyperbola $xy = c^2$, when the asymptotes are referred as the co-ordinate axes, is given by

$$xy = -c^2.$$

PROPERTIES OF CONJUGATE HYPERBOLAS

(A) If a diameter meets a hyperbola in real points, it will meet its conjugate hyperbola in imaginary points.

The equation of a hyperbola, its conjugate hyperbola and any diameter are respectively $x^2/a^2 - y^2/b^2 = 1$, ...(1)

$$x^2/a^2 - y^2/b^2 = -1, \quad ...(2)$$

and $$y = mx. \quad ...(3)$$

Putting for y from (3) in (1), the abscissae of the points of intersection of (3) with (1) are given by

$$a^2 = \frac{a^2b^2}{b^2 - a^2m^2} \quad ...(4)$$

Similarly the abscissae of the points of intersection of (3) with (2) are given by $$x^2 = \frac{-a^2b^2}{b^2 - a^2m^2} \quad ...(5)$$

If (3) meets (1) in real points, then the values of x given by (4) are real so that $b^2 - a^2m^2$ is + ive. Obviously then the values of x given by (5) are imaginary.

(B) If a pair of diameters be conjugate with respect to a hyperbola, they are also conjugate w.r.t. the conjugate hyperbola.

The diameters $x^2/a^2 - y^2/b^2 = 1$ if $m_1m_2 = b^2/a^2$. ...(1)

Also they are conjugate w.r.t. the conjugate hyperbola

$$\frac{x^2}{a^2} - \frac{y^2}{b^2} = -1 \quad \text{i.e.,} \quad \frac{x^2}{(-a^2)} - \frac{y^2}{(-b^2)} - 1$$

if $$m_1m_2 = \frac{-b^2}{-a^2} = \frac{b^2}{a^2} \text{ which is true}$$

Hence if two diameters are conjugate for a hyperbola, they are also conjugate for the conjugate hyperbola.

(C) The equations of a hyperbola and the conjugate hyperbola can always be so written that they differ from the equation of the asymptotes by the same constant.

The equation of a hyperbola, asymptotes and the conjugate hyperbola are respectively

$$x^2/a^2 - y^2/b^2 - 1 = 0, \quad ...(1)$$

$$x^2/a^2 - y^2/b^2 = 0, \quad ...(2)$$

and $x^2/a^2 - y^2/b^2 + 1 = 0.$ ...(3)

We see that if we add the constant 1 to the equation (1), we get the equation (2) and if we add the same constant 1 to the equation (2), we get the equation (3). Hence the required result is proved.

(D) A hyperbola and its conjugate hyperbola have the same pair of asymptote.

The asymptote of the hyperbola $x^2/a^2 - y^2/b^2 = 1$ are given

as $x^2/a^2 - y^2/b^2 = 0.$ ...(1)

Clearly the asymptotes of the conjugate hyperbola

$$x^2/a^2 - y^2/b^2 = -1$$

i.e., $y^2/b^2 - x^2/a^2 = 1$ are given as

$$x^2/a^2 - y^2/b^2 = 0$$

i.e., $x^2/a^2 - y^2/b^2 = 0,$ ...(2)

The equations (1) and (2) are the same. **Hence proved.**

PROPERTIES OF CONJUGATE DIAMETERS OF CONJUGATE HYPERBOLAS

(A) To prove that if a pair of conjugate diameters meet the hyperbola in P, P' and its conjugate in Q, Q', then the asymptotes bisect PQ, PQ', P'Q and P'Q'.

Let PCP' and QCQ' be a pair of conjugate diameters, the points P, P' lying on the hyperbola $x^2/a^2 - y^2/b^2 = 1$, and the points Q, Q' on the hyperbola $-x^2/a^2 - y^2/b^2 = 1$. If P is the point ($a \sec \theta$, $b \tan \theta$), then the co-ordinates of Q are ($a \tan \theta$, $b \sec \theta$).

Also then the co-ordinates of the points P' and Q' are ($-a \sec \theta$, $-b \tan \theta$) and ($-a \tan \theta$, $-b \sec \theta$) respectively.

Let (x_1, y_1) is the middle point of PQ. The we have $x_1 = \frac{1}{2}a$ $(\sec \theta + \tan \theta)$, $y_1 = \frac{1}{2}b$ $(\sec \theta + \tan \theta)$.

Dividing, we have $x_1/y_1 = a/b$ or $x_1/a = y_1/b$, showing that the point (x_1, y_1) lies on the asymptote $x/a = y/b$.

Similarly, we can show that the middle point of PQ' lies on the other asymptote $x/b = -y/b$.

Also the other two middle points can be shown to lie on the asymptote. Hence the proposition.

(B) To prove that the tangents drawn at the points where a pair of conjugate diameters meet a hyperbola and its conjugate, form a parallelogram whose vertices lie on the asymptotes and whose area is constant.

Let PCP' and QCQ' be a pair of conjugate diameters, the points P, p' being on the hyperbola $x^2/a^2 - y^2/b^2 = 1$ and the points Q, Q' on the hyperbola

$$-x^2/a^2 + y^2/b^2 = 1$$

Now if P is the point (a sec θ, b tan θ), then the co-ordinates of Q are (a tan θ, b sec θ).

Also the co-ordinates of the point P' and Q' are (–a sec θ, –b tan θ) and (–a tan θ, –b sec θ) respectively. The tangents at P and P' to the hyperbola $x^2/a^2 - y^2/b^2 = 1$ are respectively.

$$(x/a) \sec \theta - (y/b) \tan \theta = 1, \qquad ...(1)$$

and $$-(x/a) \sec \theta + (y/b) \tan \theta = 1, \qquad ...(2)$$

Again the tangents at Q and Q' to the hyperbola

$-x^2/a^2 + y^2/b^2 = 1$ are respectively

$$-(x/a) \tan \theta + (y/b) \sec \theta = 1, \qquad ...(3)$$

and $$(x/a) \tan \theta + (y/b) \sec \partial = 1, \qquad ...(4)$$

We see that the tangents (1) and (3) sec that the point of intersection k of the tangents (1) and lies on the line

$$(x/a) (\sec \theta + \tan \theta) - (y/b) \sec \theta + \tan \theta) = 0$$

i.e., $x/a - y/b = 0$, which is the asymptote KN.

Subtracting (4) from (1), we see that the point of intersection m of the tangents at P and Q' lies on the asymptote

$$x/a + y/b = 0$$

Similarly, we can show that the other two points of intersection L and N of the above tangents also lie on one or the other of the asymptotes. Hence the vertices of the parallelogram formed by these tangents lie on the asymptotes.

Now the gradient of the tangent at P is b sec θ/a tan θ which is also the gradient of the line CQ, Similarly the gradient of the tangent at Q is equal to the gradient of the line CP. Therefore the tangents at the extremities of any diameter are parallel to the conjugate diameter.

Thus C is the point of intersection of the diagonals of the parallelogram KLNM and CP, CQ are parallel to the sides. Therefore PCP' and QCQ' divide the parallelogram KLNM into four equal parallelograms. Hence the area of the parallelogram KLMN

= 4 × area of the parallelogram CQKP

= 4 × CQ × the perpendicular from C on PK whose equation is (1)

$$= 4.\sqrt{(a^2 \tan^2 \theta + b^2 \sec^2 \theta)}.\ \frac{1}{\sqrt{(\sec^2 \theta/a^2 + \tan^2 \theta/b^2)}}$$

= 4ab, which is constant.

(C) *To prove that if a pair of conjugate diameters cut two conjugate hyperbolas in P and Q respectively, then $CP^2 - CQ^2$ is constant and equal to $a^2 - b^2$ where C is the centre of the hyperbolas.*

The parametric co-ordinates of P on $x^2/a^2 - y^2/b^2 = 1$ can be taken as $(a \sec \theta_1, b \tan \theta_1)$ and of Q on $-x^2/a^2 + y^2/b^2 = 1$ as $(a \tan \theta_2, b \sec \theta_2)$.

If m_1 and m_2 are the gradients of CP and CQ respectively, then $m_1 = b \tan \theta_1/a \sec \theta_1$ and $m_2 = b \sec \theta_2/a \tan \theta_2$.

Since CP and CQ are conjugate diameters, therefore we have

$$m_1 m_2 = \frac{b^2}{a^2} \text{ i.e., } \frac{b \tan \theta_1}{a \sec \theta_1}.\frac{b \sec \theta_2}{b \tan \theta_2} = \frac{b^2}{a^2} \text{ i.e., } \frac{\sin \theta_1}{\sin \theta_2} = 1$$

i.e., $\sin \theta_1 = \sin \theta_2$.

$\therefore \quad \theta_2 = \theta_2 \quad$ or $\quad \pi = \theta_1$.

Hence $CP^2 - CQ^2 = a^2 \sec^2 \theta_1 + \theta_2 \tan^2 \theta_1 + b^2 \tan^2 \theta_1 - (a^2 \tan^2 \theta_2 + b^2 \sec^2 \theta_2)$

$= a^2 \sec^2 \theta_1 + b^2 \tan^2 \theta_1 - (a^2 \tan^2 \theta_1 + b^2 \sec^2 \theta_1)$ [by (1)]

$= a^2 (\sec^2 \theta_1 - \tan^2 \theta_1) - b^2 (\sec^2 \theta_1 - \tan^2 \theta_1) = a^2 - b^2$.

(D) *To prove that of a pair of conjugate diameters of a hyperbola one meets the hyperbola in real points, and the other meets the conjugate hyperbola in real points.*

Let the equations of a hyperbola and its conjugate hyperbola

be $\quad x^2/a^2 - y^2/b^2 = 1$. ...(1)

and $\quad -x^2/a^2 + y^2/b^2 = 1$. ...(2)

Suppose $y = m_1x$ and $y = m_2x$ are conjugate diameters of (1) and (2). Then we have $m_1m_2 = b^2/a^2$. ...(3)

The abscissae of the points of intersection of the diameter $y = m_1x$ and the hyperbola (1) are given as

$$x^2/a^2 - m_1^2x^2/b^2 = 1$$

i.e., $x^2 = a^2b^2/(b^2 - a^2m_1^2)$.

If the diameter $y = m_1x$ meets (1) at rea. points, we must have

$$b^2 - a^2m_1^2 > 0.$$

Now the abscissae of the points of intersection of the diameter $y = m_2x$ with the hyperbola (2) are given as

$$-x^2/a^2 + m_2^2x^2/b^2 = 1$$

$$\Rightarrow \quad x^2 -(b^2 + a^2m_2^2) = a\ a^2b^2$$

$$\Rightarrow \quad x^2 = \frac{a^2b^2}{-b^2 + a^2m_2^2} = \frac{a^2b^2}{-b^2 + a^2.(b^4/a^4m_1^2)} \qquad \text{[form (2)]}$$

$$\Rightarrow \quad x^2 = \frac{a^2m_1^2}{-a^2m_1^2 + b^2},$$ which given real values since $b^2 - a^2m_1^2 > 0$.

Hence, if the diameter $y = m_1x$ meets the hyperbola (1) in real points, then the conjugate diameter $y = m_2x$ meets the conjugate hyperbola (2) in real points.

MISCELLANEOUS EXAMPLES

Example 1:

The angle subtended by any chord at the centre is the supplement of the angle between the tangents at the ends of the chord.

Solution:

Let the equation of the hyperbola be $x^2 - y^2 = a^2$ and P and Q be any two points on it such that their coordinates are respectively $(a \sec \phi_1, a \tan \phi_1)$ and $(a \sec \phi_2, a \tan \phi_2)$ and C be the centre of the hyperbola.

Equation of the line PC is $y - 0 = \dfrac{a \tan\phi_1 - 0}{a \sec \phi_1 - 0}(x - 0)$

$$\Rightarrow \quad y = x \sin \phi_1 \qquad \text{...(1)}$$

Similarly equation of QC will be

$$y = x \sin \phi_2 \qquad \text{...(2)}$$

If α is the angle between PC and QC, then,

$$\tan \alpha = \frac{\sin \phi_1 - \sin \phi_2}{1 + \sin \phi_1 \sin \phi_2} \qquad \text{...(3)}$$

Again the equation to the tangent at P is

$$x.a \sec \phi_1 - y.a. \tan \phi_1 = a^2$$

$$\Rightarrow \quad y = \frac{x}{\sin \phi_1} - a\frac{\cos \phi_1}{\sin \phi_1} \quad ...(4)$$

Similarly the equation to the tangent at Q is

$$y = \frac{x}{\sin \phi_2} - a\frac{\cos \phi_2}{\sin \phi_2} \quad ...(5)$$

If β is the angle between the tangents at P and Q, then,

$$\tan \beta = \frac{\frac{1}{\sin \phi_1} - \frac{1}{\sin \phi_2}}{1 + \frac{1}{\sin \phi_1 \sin \phi_2}} = \frac{\sin \phi_2 - \sin \phi_1}{1 + \sin \phi_1 \sin \phi_2}$$

$$= -\left(\frac{\sin \phi_1 - \sin \phi_2}{1 + \sin \phi_1 \sin \phi_2}\right) = -\tan \alpha \quad \text{by (3)}$$

$$\Rightarrow \quad \tan \beta = \tan (\pi - \alpha)$$

Hence $\beta = \pi - \alpha$.

Example 2:

Show that the locus of the centre of a circle which touches externally two given circles is a hyperbola.

Solution:

Let A and B the centres of the fixed circles whose radii are respectively r_1 and r_2. Taking AB as the axis of x and the right bisector of AB as y-axis. The coordinates of A and B may be given by (–a, 0) and (a, 0). Let C be the centre of the variable circle having coordinates as (h, k) and radius r. If the circles touch the first circle i.e. having the centre at A, externally, clearly the distance between the centres must be equal to the sum of the radii.

Hence $CA = r_1 = r$

$$\Rightarrow \quad \sqrt{[(h+a)^2} + (k - 0)^2] = r_1 = r \quad ...(1)$$

Similarly, if the variable circle touches the circle with centre at B and radius r_2, externally, we must have

$$CB = r_2 + r$$

$$\Rightarrow \quad \sqrt{[(h-a)^2} + (k - 0)^2]\ r_2 + r \quad ...(2)$$

Subtracting (2) from (1), we get

$$\sqrt{[(h+a)^2+k^2]}-\sqrt{[(h-a)^2+k^2]}=r_1-r_2$$

$$\Rightarrow \sqrt{[(h+a)^2+k^2]}-(r_1-r_2)+\sqrt{(h-a)^2+k^2]}$$

Squaring we get

$$h^2 + a^2 + 2ah + k^2 = (r_1 - r_2)^2 + h^2 + k^2 - 2ah + 2(r_1 - r_2)\sqrt{[(h-a)^2+k^2]}$$

$$\Rightarrow 4ah - (r_1 - r_2)^2 = 2(r_1 - r_2)\sqrt{[(h-a)^2+k^2]}$$

Squaring again we get,

$$16a^2h^2 + (r_1 - r_2) - 8ah(r_1 - r_2)^2 = 4(r_1 - r_2)^2[h^2 + a^2 - 2ah + k^2]$$

$$\Rightarrow \quad 4(r_1 - r_2)^2k^2 - h^2[16a^2 - 4(r_1 - r_2)^2] = (r_1 - r_2)^4$$

$$\Rightarrow \quad \frac{k^2}{\left[\frac{r_1-r_2}{2}\right]^2}-\frac{h^2}{\left[(r_1-r_2)^2/16a^2-4(r_1-r_2)^2\right]}=1$$

Generalising for (h, k), we get the required locus as

$$\frac{y^2}{\left[\frac{r_1-r_2}{2}\right]^2}-\frac{x^2}{\left[(r_1-r_2)^2/16a^2-4(r_1-r_2)^2\right]}=1$$

which is clearly is hyperbola.

Example 3(a):

Find the points common to the hyperbola $25x^2 - 9y^2 = 225$ and the straight line $25x + 12y - 45 = 0$.

Solution:

The equation to the hyperbola is given as

$$25x^2 - 9y^2 = 225 \quad ...(1)$$

and that of the line as $25x + 12y - 45 = 0$ ...(2)

To get the common points of the two curves i.e. the points of intersection of (1) and (2), we have to solve them simultaneously. Hence by (2)

$$y=\frac{45-25x}{12} \quad ...(3)$$

Putting the value of y in (1), we get

$$25x^2 - 9 \times \left(\frac{45-25x}{12}\right)^2 = 225$$

$$\Rightarrow \quad 25x^2 - 9\left(\frac{2025+625x^2-2250x}{144}\right) = 225$$

$$\Rightarrow \quad 400x^2 - 2025 - 625x^2 + 2250x - 3600 = 0$$

$$\Rightarrow \quad -225x^2 + 2250x - 5625 = 0$$

$$\Rightarrow \quad x^2 - 10x + 25 = 0$$

$$\Rightarrow \quad (x-5)^2 = 0 \text{ whence } x = 5.$$

Hence by (3), we get

$$y = \frac{45-25\times5}{12} = -\frac{20}{3}$$

Hence there is only one common point of (1) and (2) and its coordinates are (5, – 20/3).

Example 3(b):

In the hyperbola $4x^2 - 9y^2 = 36$, find the axes, the coordinates of the foci, the eccentricity and the latus rectum.

Solution:

The equation to the hyperbola is given as

$$4x^2 - 9y^2 = 36$$

$$\Rightarrow \quad \frac{x^2}{9} - \frac{y}{4} = 1 \qquad ...(1)$$

Comparing with the standard equation

$$\frac{x^2}{a^2} - \frac{y^2}{b^2} = 1 \qquad ...(2)$$

We find $a^2 = 9$, $b^2 = 4$ where $a = 3$, $b = 2$

Hence the transverse axis = 2a = 2 × 3 = 6

and the conjugate axis = 2b = 2 × 2 = 4

Again $e = \sqrt{1+\frac{b^2}{a^2}} = \sqrt{1+\frac{4}{9}} = \frac{\sqrt{13}}{9}$

The coordinates of the foci are (± ae, 0), hence the foci of the given hyperbola are

$$\left(3\frac{\sqrt{13}}{3}, 0\right) \text{ and } \left(-3\frac{\sqrt{13}}{3}, 0\right)$$

i.e. $\left(\sqrt{13}, 0\right)$ and $\left(-\sqrt{13}, 0\right)$

Latus rectum $= 2.\dfrac{b^2}{a} = 2.\dfrac{4}{3} = \dfrac{8}{3}$.

Example 3(c):

Find the equation to the hyperbola of given transverse axis whose vertex bisects the distance between the centre and the focus.

Solution:

Let the equation to the hyperbola be

$$\frac{x^2}{a^2} - \frac{y^2}{b^2} = 1 \qquad ...(1)$$

Hence the transverse axis is 2a. If c is the centre, S is the focus and A be the vertex to the hyperbola.

CS = ae (distance between the centre and the focus)

Hence by hypothesis $a = \dfrac{ae}{2}$ or $e = 2$

Again $b^2 = a^2 (e^2 - 1) = a^2 (4 - 1) = 3a^2$

Substituting the value of b^2 in (1), we get

$$\frac{x^2}{a^2} - \frac{y}{3a^2} = 1$$

$\Rightarrow$ $3x^2 - y^2 = 3a^2$. **Ans.**

Example 3(d):

Find the equation of the tangent to the hyperbola $4x^2 - 9y^2 = 1$ which is parallel to the line $4y = 5x + 7$.

Solution:

The equation to the hyperbola is given as

$$4x^2 - 9y^2 = 1 \quad \text{or} \quad \frac{x^2}{1/4} - \frac{y^2}{1/9} = 1 \qquad ...(1)$$

and the line is given as $4y = 5x + 7$ ...(2)

Comparing (1) with the standard equation of hyperbola

$$\frac{x^2}{a^2}-\frac{y^2}{b^2}=1$$

$$a^2=\frac{1}{4} \text{ and } b^2=\frac{1}{9}$$

The equation of any tangent to the hyperbola is given by

$$y = mx \pm \sqrt{(a^2m^2 - b^2)} \qquad ...(3)$$

If (3) is parallel to (2), then equating the slopes, we get,

$$m = 5/4$$

Substituting the values of a^2, b^2 and m in (4) the equation tot he tangent is

$$y = y=\frac{5}{4}x \pm \sqrt{\frac{1}{4}\left(\frac{5}{4}\right)^2-\frac{1}{9}}=\frac{5}{4}x \pm \sqrt{\frac{225-64}{64\times 9}}$$

$$\Rightarrow \quad y=\frac{5}{4}x\pm\frac{1}{8.3}\sqrt{161} \quad 24y = 30x \pm \sqrt{161}$$

$$\Rightarrow \quad 24y - 30x = \pm \sqrt{161}.$$ **Ans.**

Example 4:

Prove that a circle can be drawn through the foci of a hyperbola and the points in which any tangent meets the tangents at the vertices.

Solution:

Let the equation of the hyperbola be

$$\frac{x^2}{a^2}-\frac{y^2}{b^2}=1 \qquad ...(1)$$

and let A and A' be it vertices, S and S' be its foci, LM and L'M' be the tangents at the vertices A and A' respectively and P be any point on the hyperbola having the coordinates as (a cos ϕ, b tan ϕ). If tangent to (1) at P cuts the tangents at the vertices i.e. LM and L'M' at Q and R respectively, then the equation to the tangent PQR will be

$$\frac{x}{a}\sec\phi-\frac{y}{b}\tan\phi=1$$

or $$\frac{x}{a}-\frac{y}{b}\sin\phi=\cos\phi \qquad ...(2)$$

The equation to the tangent at A i.e. (a, 0) is

$$x = a \qquad ...(3)$$

Solving (2) and (3) we get the coordinates of the point of intersection

$$Q \text{ as } \left\{a, \frac{b\left(1-\cos\theta\right)}{\sin\theta}\right\} \text{ i.e. } (a,\ b \tan \phi/2)$$

Again the equation to the tangent at the other vertex $A' \equiv (-a, 0)$ will be

$$x = -a \qquad ...(4)$$

Solving (2) and (4) the coordinates of R will be

$$(-a,\ -b \cot - \phi/2)$$

Now, we are to prove that all the four points S, S', Q and R lie on a circle. Let the circle passing through S, S' and Q be

$$x^2 + y^2 + 2gx + 2fy + c = 0 \qquad ...(5)$$

The coordinates of S, S' and Q are respectively (ae, 0), (– ae, 0) and (a, b tan ϕ/2). As all of these points lie on (5). These coordinates will satisfy it. Taking these points one by one, we get

$$a^2e^2 + 0\ 2gae + 0 + c = 0 \qquad ...(6)$$

$$a^2e^2 + 0 - 2gae + 0 + c = 0 \qquad ...(7)$$

and $$a^2 + b^2 \tan^2 \phi/2 + 2ag + 2fb \tan \phi/2 + c = 0 \qquad ...(8)$$

Solving (6) and (7) by subtraction and addition, we get

$$g = 0 \text{ and } c = -a^2e^2$$

Putting these values in (8), we get

$$a^2 + b^2 \tan^2 \phi/2 + 2fb \tan \phi/2 - a^2 e^2 = 0$$

$$\Rightarrow b^2 \tan^2 \phi/2 + 2\ fb \tan \phi/2 = a^2 = a^2 (e^2 - 1) = a^2 \left\{\left(1+\frac{b^2}{a^2}\right)-1\right\}$$

$$= b^2 \quad \therefore \quad e^2 = \left(1+\frac{b^2}{a^2}\right)$$

$$\Rightarrow \quad 2fb \tan \phi/2 = b^2 - b^2 \tan^2 \phi/2 = b^2 (1 - \tan^2 \phi/2)$$

$$= b^2 \left(\frac{\cos^2 \phi/2 - \sin^2 \phi/2}{\cos^2 \phi/2}\right)$$

$$\Rightarrow \quad 2fb \frac{\sin \phi/2}{\cos \phi/2} = b^2 \left(\frac{\cos^2 \phi/2 - \sin^2 \phi/2}{\cos^2 \phi/2}\right)$$

Hence $f = \frac{b}{2}\left(\frac{\cos^2 \phi/2 - \sin^2 \phi/2}{\sin \phi/2.\cos \phi/2}\right) = \frac{b}{2}$ (cot ϕ/2 – tan ϕ/2)

Substituting the values of g, f and c is (5) the equation of the circle passing through S, S' and P is

$$x^2 + y^2 + 2\frac{b}{2}\left(\cot\frac{\phi}{2} - \tan\frac{\phi}{2}\right) y - a^2 e^2 = 0$$

$$\Rightarrow \quad x^2 + y^2 + b\left(\cot\frac{\phi}{2} - \tan\frac{\phi}{2}\right) y - a^2 e^2 = 0 \qquad ...(9)$$

If the point R (– a, – b cot ϕ/2) lies on (9), then its coordinates may satisfy (9). Hence putting the coordinates, we get the L.H.S. of (9) as

$$a^2 + b^2 \cot^2\frac{\phi}{2} + b\left(\cot\frac{\phi}{2} - \tan\frac{\phi}{2}\right)\left(-b\cot\frac{\phi}{2}\right) - a^2 e^2$$

$$= a^2 + b^2 \cot^2\frac{\phi}{2} + b^2 \cot\frac{\phi}{2} + b^2 - a^2\left(1 + \frac{b^2}{a^2}\right)$$

$= 0 =$ R.H.S. of (9)

Hence P lies on the circle passing through S, S' and Q. **Proved.**

Example 5:

In a rectangular hyperbola, prove that

SP.SP' = CP^2

Solution:

Take any point P whose coordinates are (a sec f, a tan f) on the rectangular hyperbola, having its foci as S and S' and the centre at C.

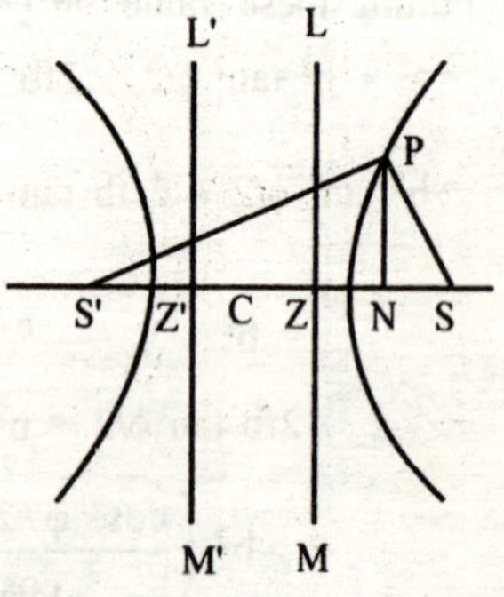

If Z be the point of intersection of one of the directrixes with the perpendicular from P on the transverse axis, then,

SP = e.NZ = e (CN – CZ)

$$= e\left(a \sec\phi - \frac{a}{e}\right)$$

$$\Rightarrow \quad SP = ae \sec\phi - a \qquad ...(1)$$

Again S'P = eNZ' = e (Z'C + CN)

$$= e\left(\frac{a}{e} + a \sec \phi\right)$$

$\Rightarrow$ S'P = a + ae sec ϕ ...(2)

Multiplying (1) and (2), we get

$$SP\ S'P = (ea \sec \phi - a)(ea \sec \phi + a)$$
$$= e^2a^2 \sec^2 \phi - a^2$$
$$= a^2 (e^2 \sec^2 \phi - 1)$$

For the rectangular hyperbola, $e^2 = 2$, so putting this value we get,

$$SP.S'P = a^2 (2 \sec^2 \phi - 1)\ a^2 (\sec^2 \phi + \sec^2 - 1)$$
$$= a^2 \sec^2 \phi + a^2 \tan^2 \phi = CP^2.$$ **Proved.**

Example 6:

In the hyperbola $16x^2 - 9y^2 = 144$, find the equation to the diameter which is conjugate to the diameter whose equation is $x = 2y$.

Solution:

The given equation is

$$16x^2 - 9y^2 = 144$$

$$\Rightarrow \quad \frac{x^2}{9} - \frac{y^2}{16} = 1.$$

Hence comparing with the standard equation of hyperbola, we have $a^2 = 9$ and $b^2 = 16$. The given diameter is x = 2y or y = 1/2 x. Hence of y = mx is the conjugate diameter, then

$$\frac{1}{2}m = \frac{4^2}{3^2} \text{ or } m = \frac{12}{9}.$$

Putting the value of m in y = mx, we get the required conjugate diameter

y = (32/9)x

or 9y = 32 x. **Ans.**

Example 7(a):

In the rectangular hyperbola the distance of any point from the centre varies inversely as the perpendicular from the centre upon its polar.

Solution:

Let the equation of the rectangular hyperbola be

$$x^2 - y^2 = a^2$$

If (x_1, y_1) be any point on it, then its distance from the centre will be $\sqrt{(x_1^2 + y_1^2)}$.

The equation to the polar of the point (x_1, y_1) with respect to the hyperbola will be,

$$xx_1 - yy_1 = a^2$$

The length of the perpendicular from the origin upon the polar will be

$$a^2 . \frac{1}{\sqrt{x_1^2 + y_1^2}}$$

This clearly varies inversely as $\sqrt{(x_1^2 + y_1^2)}$, which is the distance of the point from the origin i.e. the centre.

Example 7(b):

Find the equation to the hyperbola, whose eccentricity is 5/4, where focus is (a, 0) and whose directrix is 4x – 3y = a.

Find also the coordinates of the centre and the equation to the other directrix.

Solution:

By definition, the equation to the hyperbola having the focus (a, 0) and directrix as

$$4x - 3y = a \qquad ...(1)$$

and eccentricity as 5/4, will be

$$(x - a)^2 + (y - 0)^2 = \left(\frac{5}{4}\right)^2 \left\{\frac{4x - 3y - a}{\sqrt{(4^2 + 3^2)}}\right\}^2$$

$$\Rightarrow x^2 - 2ax + a^2 + y^2 = \frac{15}{16} \times \frac{15x^2 + 9y^2 + a^2 - 24xy - 8ax + 6ay}{25}$$

$$\Rightarrow (x^2 - 2x + a^2 + y^2) \times 16 = 16x^2 + 9y^2 + a^2 - 24xy - 8ax + 6ay$$

$$\Rightarrow 7y^2 + 24xy - 24ax - 6ay + 15a^2 = 0$$

This is the equation to the required hyperbola.

Let the given focus (a, 0) be S. The axis will be a line passing through (a, 0) and perpendicular to the directrix

$$4x - 3y = a$$

Any line perpendicular to directrix is

$$3x + 4y = \lambda \text{ (where } \lambda \text{ is any constt.)}$$

As the line passes through (a, 0), we have

$$3a + 0 = \lambda \quad \therefore \quad \lambda = 3a$$

Hence the equation to the axis is

$$3x + 4y = 3a \qquad ...(2)$$

Let Z be the point of intersection of the axis (2) with the directrix (1), then the coordinates of Z will be obtained by solving (2) and (1). Solving, we get these coordinates as

$$\left(\frac{13a}{25}, \frac{9a}{25}\right)$$

The centre divides the join of S and Z externally in the ratio

$$1: e^2 \text{ or } 1: (5/4)^2 \text{ or } 16: 25.$$

Hence the coordinates of the centre C will be

$$\left(\frac{16\times a-\frac{13a}{25}\times 25}{16-25}, \frac{16\times 0-\frac{25\times 9a}{25}}{16-25}\right)$$

$$\Rightarrow \quad \frac{3a}{-9} \text{ and } \frac{-9a}{-9} \text{ or } \left(-\frac{a}{3}, a\right)$$

Let the other directrix intersect the axis at Z' and (x, y) be its coordinates, then

$$x+\frac{13a}{25}=-\frac{2a}{3} \text{ and } y+\frac{9a}{25}=2a$$

$$\Rightarrow \quad x=-\frac{13a}{25}-\frac{2a}{3} \text{ and } y=2a-\frac{9a}{25}$$

$$\Rightarrow \quad x=\frac{-39a-50a}{75}=\frac{-89a}{75} \text{ and } y=\frac{41a}{25}$$

Therefore, the other directrix is the line passing through the point $\left(-\frac{89}{25}a, \frac{41}{25}a\right)$ and parallel to the given directrix (1).

The equation to the line parallel to (1) may be written as

$$4x - 3y = \lambda \qquad ...(3)$$

As (3) passes through $\left(-\frac{889}{775}a, \frac{41}{25}a\right)$, we have,

$$-4 \times \frac{889a}{775} - 3 \times \frac{41a}{25} = \lambda$$

or $$\lambda = -\frac{229}{33}a$$

Substituting the value of l in (3), we get the equation to the other directrix as

$$4x - 3y = -\frac{229}{33}a$$

$$\Rightarrow \quad 12x - 9y + 229a = 0.$$

Example 8:

In both an ellipse and a hyperbola, prove that the focal distance of any point and the perpendicular from the centre upon the tangent at it meet on a circle whose centre is the focus and where radius is the semitransverse axis.

Solution:

Let the equation to any hyperbola be given as

$$\frac{x^2}{a^2} - \frac{y^2}{b^2} = 1$$

and 'P' be any point on it as $(a \sec \phi, b \tan \phi)$. The focus S will be clearly $(ae, 0)$.

Equation to the focal chord PS will be

$$y - 0 = \frac{b \tan \phi - 0}{a \sec \phi - ae}(x - ae)$$

$$\Rightarrow \quad y = \frac{(x - ae)\, b \sin \phi}{a - ae \cos \phi} \qquad ...(1)$$

Again the equation to the tangent at the point P to the hyperbola will be

$$\frac{x}{a} - \frac{y}{b} \sin \phi = \cos \phi \qquad ...(2)$$

Any line passing through the centre (0, 0) may be given by

$$y = mx \qquad ...(3)$$

If (3) is perpendicular to (2), then we have

$$m \times \frac{1}{a} \cdot \frac{b}{\sin \phi} = -1$$

or $$m = -\frac{a \sin \phi}{b}$$

Hence the equation to the line perpendicular to the tangent at P from the centre C will be

$$y = -\frac{a}{b} \sin \phi . x$$

$\Rightarrow$ $$by + ax \sin \phi = 0 \qquad ...(4)$$

To find out the locus of the point of intersection of (2) and (1), we have to eliminate the variable 'ϕ' between them

By (4) $$\sin \phi = \frac{y\,b}{a\,x} \qquad ...(5)$$

Now putting the value of sin ϕ from (5) in (1)

$$ay - ae\, y \cos \phi = -(x - ae)\, b \times \frac{yb}{xa}$$

Hence $$\cos \phi = \frac{a^2 x + (x - ae) b^2}{a^2 ex} = \frac{x\left(a^2 + b^2\right) - aeb^2}{a^2 ex}$$

$$= \frac{xa^2 e^2 - aeb^2}{a^2 ex} \left(\text{as } a^2 + b^2 = a^2 e^2\right)$$

$\Rightarrow$ $$\cos \phi = \frac{xae - b^2}{ax} \qquad ...(6)$$

Squaring and adding (5) and (6), we get

$$\frac{y^2 b^2}{a^2 x^2} + \frac{\left(xae - b^2\right)^2}{a^2 x^2} = 1$$

$\Rightarrow$ $$y^2b^2 + (xae - b^2)^2 = a^2x^2.$$

Simplifying and putting $a^2 + b^2 = a^2e^2$, we get the required locus as $(x - ae)^2 + y^2 = a^2$, which is clearly a circle having its centre at (ae, 0), that is the focus and its radius is a, i.e. the semi transverse axis.

Similarly taking the equation to the ellipse as $(x^2/a^2) + (y^2/b^2) = 1$, any point on it as (a cos ϕ, b sin ϕ) and focus as (ae, 0) and using the relation $a^2 - b^2 = a^2e^2$, we can prove the same result.

Example 9:

Find the equation to, and the length of the common tangent to the two hyperbolas

$$\frac{x^2}{a^2}-\frac{y^2}{b^2}=1 \text{ and } \frac{y^2}{a^2}-\frac{x^2}{b^2}=1$$

Solution:

The equation to the hyperbolas are given as

$$\frac{x^2}{a^2}-\frac{y^2}{b^2}=1 \qquad ...(1)$$

and
$$\frac{y^2}{a^2}-\frac{x^2}{b^2}=1 \qquad ...(2)$$

The equation to the tangent at any point $P \equiv (a \sec \phi, b \tan \phi)$ on (1) to the same hyperbola is

$$\frac{x}{a}-\frac{y}{b}\sin\phi=\cos\phi$$

$$\Rightarrow \quad bx - ay \sin\phi = ab\cos\phi \qquad ...(3)$$

If (3) is the tangent to the hyperbola (2), then we must have,

$$a^2b^2\cos^2\phi = -b^4 + a^4\sin^2\phi$$

$$\Rightarrow \quad a^2b^2 + b^4 = (a^4 - b^4)\tan^2\phi$$

$$\Rightarrow \quad b^2(a^2+b^2) = (a^2+b^2)(a^2-b^2)\tan^2\phi$$

$$\Rightarrow \quad b^2 = (a^2-b^2)\tan^2\phi$$

Hence $\tan^2\phi = \dfrac{b^2}{a^2-b^2}$

or $\tan\phi = \dfrac{b}{\sqrt{a^2-b^2}}$

Hence $\sin\phi = \pm b/a$

and $\cos\phi = \pm \dfrac{1}{a}\sqrt{a^2-b^2}$

Putting these values in (3), the conjugate diameters are

$$bx \pm ay = ab \,.\, \frac{1}{a}\sqrt{a^2-b^2}$$

or
$$y = \pm x \pm \sqrt{a^2-b^2}$$

If ϕ is less than $\pi/2$, the both $\sin\phi$ and $\cos\phi$ will be positive. So taking $\phi < \pi/2$ and putting the values of $\sin\phi$ and $\cos\phi$ is (3), we get,

$$bx - ay \cdot \frac{b}{a} = \frac{ab\sqrt{a^2 - b^2}}{a}$$

$$\Rightarrow \qquad y = x - \sqrt{a^2 - b^2} \qquad \text{...(6)}$$

The point where (6) touches the hyperbola (2) is obtained by solving the equation of common tangent i.e. (6) and the hyperbola (2).

Solving, we get the point say Q, as

$$\left(-\frac{b^2}{\sqrt{a^2 - b^2}}, \frac{a^2}{\sqrt{a^2 - b^2}}\right)$$

The point (a sec f, b tan f) on (1) by putting the values of $\sec\phi$ and tan f by relations (4) and (5) becomes,

$$\left(\frac{a^2}{\sqrt{a^2 - b^2}}, \frac{b^2}{\sqrt{a^2 - b^2}}\right)$$

The required length of the common tangent will be distance between the two points of contact, i.e. P and Q

So $$PQ = \left(a^2 + b^2\right)\sqrt{\frac{2}{\sqrt{a^2 - b^2}}}.$$

Example 10(a):

If in the rectangular hyperbola the normal at P meet the axes is G and g, then

$$PG = Pg = PC.$$

Solution:

Let the equation to the rectangular hyperbola be,

$$x^2 - y^2 = a^2 \qquad \text{...(1)}$$

If P be any point on it as $(a\sec\phi, a\tan\phi)$, then the equation to the normal at P will be given by

$$x\sin\phi + y = 2a\tan\phi$$

Its intersection with $y = 0$ gives $x = 2a$ sec f. As the point of intersection is G and C, the centre of the hyperbola hence

$$CG = 2a\sec\phi$$

Again, the intersection of the normal with $x = 0$ gives $y = 2a\tan\phi$. If g denote this point of intersection, we have

$Cg = 2a \tan \phi$

Now $PG^2 = (a \sec \phi - 2a \sec \phi)^2 + (a \tan \phi - 0)^2$

$a^2 (\sec^2 \phi + \tan^2 \phi)$

$Pg^2 = (a \sec f - 0)^2 + (a \tan f - 2a \tan f)^2$

$= a^2 (\sec^2 \phi + \tan^2 \phi)$

$PC^2 = (a \sec f - 0)^2 + (a \tan f - 0)^2$

$= a^2 (\sec^2 \phi + \tan^2 \phi)$

Hence $PG = Pg = PC$. **Proved.**

Example 10(b):

If the ordinate MP of a hyperbola be produced to Q so that MQ is equal to either of the focal distances of P, prove that the locus of Q is one or other of a pair of parallel straight lines.

Solution:

Let the equation to the hyperbola bc $\frac{x^2}{a^2} - \frac{y^2}{b^2} = 1$ and P be any point on it such that its coordinates are $(a \sec \phi, b \tan \phi)$. Drop PM perpendicular from P on axis of x so that MP is the ordinate. If S and S' be the foci, then,

$SP = ae \sec \phi - a$

and $S'P = ae \sec \phi + a$

(i) Produce MP to Q such that MQ = SP, then if the coordinates of Q be (x, y) then clearly

$x = a \sec \phi$ ...(1)

and $y = (ae \sec \phi - a)$ ...(2)

By (4) and (5) we get,

$y = ex + a$

Hence the locus of Q is either $y = ex - a$ or $y = cx + a$ each of which represents systems of parallel lines.

Example 11:

If one axis of a varying central conic be fixed in magnitude and position, prove that the locus of point of contact of a tangent drawn to it from a fixed point on the other axis is a parabola.

Solution:

Take the transverse axis as the fixed one in magnitude and position and the conjugate axis as variable.

Let $\frac{x}{a} - \frac{y}{b} \sin \phi = \cos \phi$ ...(1)

be any tangent to the hyperbola at the point (a sec f, b tan f) (when the curve is taken as hyperbola).

If (1) cuts the axis of y i.e. x = 0 at a point P, then the coordinates of P will be obtained by putting x = 0 in (1).

Hence for P, the coordinate is

$$y = -b \cot \phi$$

This point 'P' is the fixed point, hence

$$b \cot \phi = \text{constant} = \lambda \text{ (say)}$$

Let the point (a sec ϕ, b tan ϕ) (when the curve is taken as hyperbola).

If (1) cuts the axis of y i.e. x = 0 at a point P, then the coordinate of P will be obtained by putting x = 0 in (1).

Hence for P, the coordinate is

$$y = -b \cot \phi$$

This point 'P' is the fixed point, hence

$$b \cot \phi = \text{constant} = \lambda \text{ (say)}$$

Let the point (a sec ϕ, b tan ϕ), where the tangent given by (1) touches the hyperbola, be Q, hence for Q.

the abscissa – x = a sec ϕ ...(2)

and the ordinate = y = b tan ϕ ...(3)

To find the locus of Q, we have to eliminate the variables b and ϕ from (1), (2) and (3).

By (1) $b = \lambda \tan \phi$, putting in (3), we get

$$y = \tan \phi . \lambda \tan \phi = \lambda \tan^2 \phi$$

or $\tan^2 \phi = \frac{y}{\lambda}$ ...(4)

Squaring (2) we get,

$$\sec^2\phi = \frac{x^2}{a^2} \qquad ...(5)$$

Subtracting (4) from (5), we get

$$\frac{x^2}{a^2} - \frac{y}{\lambda} = \sec^2 \phi - \tan^2 \phi = 1$$

which is the required locus and is clearly a parabola.

Similarly if the curve be an ellipse, taking its equation as

$$\frac{x^2}{a^2}+\frac{y^2}{b^2}=1$$

the major axis being fixed, and any point as ($a\cos\phi$, $b\sin\phi$) it may be proved that the required locus is again a parabola.

Example 12:

The angles subtended at its vertices by any chord which is parallel to its conjugate axis is supplementary.

Solution:

Let the equation to the hyperbola be

$$x^2-y^2=a^2$$

If P be any point on it having its coordinates as ($a\sec\phi$, $a\tan\phi$) and Q be the other point of the chord, suppose that PQ is parallel to the conjugate axis, i.e. y axis. Its coordinates may be taken as ($a\sec\phi$, $-a\tan\phi$).

If A and A' be the vertices of the hyperbola, then their coordinates will be (a, 0) and (–a, 0) respectively.

The equation to the line PA will be

$$y-0=\frac{a\tan\phi-0}{a\sec\phi-a}(x-a)$$

$$\Rightarrow \quad y=\frac{(x-a)\tan\phi}{(\sec\phi-1)}$$

$$=x\frac{\tan\phi}{\sec\phi-1}-\frac{a\tan\phi}{\sec\phi-a} \qquad ...(1)$$

If this line makes an angle α with the axis of x, then

$$\tan\alpha=\frac{\sin\phi}{1-\cos\phi} \qquad ...(2)$$

Similarly if the line PA' makes an angle β with the axis of x, then we have

$$\tan\beta=\frac{\sin\phi}{1+\cos\phi} \qquad ...(3)$$

Hence $\tan\alpha.\tan\beta=\dfrac{\sin\phi}{1-\cos\phi}.\dfrac{\sin\phi}{1+\cos\phi}$ from (2) and (3)

$\Rightarrow \qquad \tan \alpha . \tan \beta = 1$

$$\Rightarrow \qquad \tan \alpha = \frac{1}{\tan \beta} = \cot \beta = \tan\left(\frac{\pi}{2} - \beta\right)$$

$\Rightarrow \qquad \alpha = 90 - \beta$ or $\alpha + \beta = 90^o$ ...(4)

Similarly if the lines QA and QA' make angles a' and b' with x-axis, be symmetry we can say

$\alpha' + \beta' = 90^o$ ...(5)

by (4) and (5) $\alpha + \alpha + \beta + \beta' = 180^o$

$\Rightarrow \qquad \angle PAO + \angle PAQ' = 180^o$. **Hence proved.**

Example 13:

On a level plain the crack of the rifle and the thud of the ball striking the target are heard at the same instant, prove that the locus of the hearer is a hyperbola.

Solution:

Suppose A to be the target and B to be the firing point and let the hearer be at a point P.

If v_1 and v_2 are the velocities of sound and the bullet respectively, then the time taken by the bullet in reaching from B to the target A = (BA/v_2) and the time taken by the sound in reaching from A to P = (AP/v_1).

Again, the time taken by the sound in reaching from B to P = (BP/v_1).

As by the hypothesis, the sound reaches the hearer i.e. at P, simultaneously, the time taken by the sound to reach up to the position P from the target together with the time taken by the bullet to reach the target must be same as the time taken by the sound in reaching from the rifle up to P.

Hence $$\frac{BA}{v_2} + \frac{AP}{v_1} = \frac{BP}{v_1}$$

$$\Rightarrow \qquad \frac{BA}{v_2} = \frac{BP}{v_1} - \frac{AP}{v_1} = \frac{1}{v_1}(BP - AP)$$

$$\Rightarrow \qquad BP - AP = \frac{v_1}{v_2} AB.$$

As v_1, v_2 and AB are constant, hence

$BP - AP = $ constt.

Therefore, the locus of the point P is a hyperbola having focii at A and B transverse axis equal to $\frac{v_1}{v_2}.AB$. **Hence Proved.**

Example 14:

Given the base of a triangle and the ratio of the tangents of half the base of angles, prove that the vertex moves on a hyperbola whose foci are extremities of the base.

Solution:

If ABC be any triangle whose sides BC, CA and AB be respectively a, b and c and S be equal to 1/2 (a + b + c) and r be the radius of in circle of the triangle ABC, then

$$\tan\frac{A}{2}=\frac{r}{S-a},\ \tan\frac{B}{2}=\frac{r}{S-b}$$

and $$\tan\frac{C}{2}=\frac{1}{S-c}.$$

Suppose AB to the base and C be the moving point, then A and B are the base angles and by hypothesis

$$\frac{\tan(A/2)}{\tan(B/2)}=\frac{r/(S-a)}{r/(S-b)}=\frac{S-b}{S-a}=\text{constt.} - \lambda \text{ (say)}$$

$$\Rightarrow \quad \lambda=\frac{\frac{a+b+c}{2}-a}{\frac{a+b+c}{2}-b}=\frac{b+c-a}{c+a-b}$$

$$\Rightarrow \quad \frac{\lambda}{1}=\frac{b+c-a}{c+a-b}$$

Hence by componando and dividando,

$$\frac{1+\lambda}{1-\lambda}=\frac{2c}{2(b-a)}=\frac{c}{b-a}$$

$$\Rightarrow \quad b-a\equiv\frac{c(1-\lambda)}{1+\lambda}$$

As c and l are constant hence (b – a) = constant, and therefore, the locus of the vertex C is a hyperbola.

Example 15:

Show that the locus of poles with respect to the parabola $y^2 = 4ax$ of tangents to the hyperbola $x^2 - y^2 = a^2$ is the ellipse $4x^2 + y^2 = 4a^2$.

Solution:

The equation of hyperbola is given as

$$x^2 - y^2 = a^2 \qquad ...(1)$$

and that of the parabola is given as

$$y^2 = 4ax \qquad ...(2)$$

The equation to any tangent to the hyperbola will be

$$x - y \sin\phi - a\cos\phi = 0 \qquad ...(3)$$

Let pole of (3) with respect to (2) be the point (h, k), then the equation to the polar of the point (h, k) with respect to the parabola will be

$$yk = 2a(x + h)$$

$$\Rightarrow \quad 2ax - yk + 2ah = 0 \qquad ...(4)$$

According to the given condition (3) and (4) must be identical hence comparing the coefficient, we get

$$\frac{2a}{1} = \frac{k}{\sin\phi} = \frac{-2ah}{a\cos\phi}$$

whence $\sin\phi = \dfrac{k}{2a}$ ...(5)

and $\cos\phi = -\dfrac{h}{a}$ ...(6)

The required locus will be obtained by eliminating ϕ between (5) and (6). By plane trigonometry

$$\sin^2\phi + \cos^2\phi = 1$$

Putting the values from (5) and (6), we get

$$\frac{k^2}{4a^2} + \frac{h^2}{a^2} = 1$$

$$\Rightarrow \quad k^2 + 4h^2 = 4a^2$$

Generalising for (h, k), we get the required locus as

$$4x^2 + y^2 = 4a^2.$$

Proved.

Example 16:

The normal to the hyperbola $\dfrac{x^2}{a^2} - \dfrac{y^2}{b^2} = 1$ *meets the axes on M and N, and lines MP and NP are drawn at right angles to the axes.*

Prove that the locus of P is the hyperbola

$$a^2x^2 - b^2y^2 = (a^2 + b^2)^2.$$

Solution:

Let the equation to the hyperbola; be

$$\frac{x^2}{a^2}-\frac{y^2}{b^2}=1$$

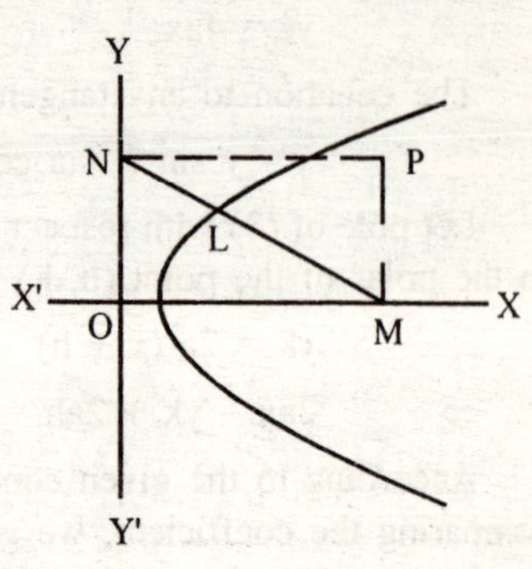

and L be any point on it having the coordinates (a sec ϕ, b tan ϕ); then the equation to the normal at this point will be given by

$$ax \sin \phi + by = (a^2 + b^2) \tan\phi \quad ... (1).$$

Let this normal cut the axis of x at M whose coordinates are (x, 0) and the axis of y at N whose coordinates are (0, y).

Solving (1) with axis of x i.e. y = 0, we get

$$x=\frac{a^2+b^2}{a\cos\phi} \qquad ...(2)$$

Similarly solving (1) with the axis of y i.e. x = 0, we get

$$y=\frac{\left(a^2+b^2\right)\tan\phi}{b} \qquad ...(3)$$

If PM and PN be the lines parallel to the axes, the coordinates of P ≡ (x, y) will be clearly gives by (2) and (3).

The required locus of P will be obtained by eliminating ϕ from (2) and (3).

From (2)

$$\cos\phi=\frac{a^2+b^2}{ax} \quad \text{hence} \quad \tan\phi=\frac{\sqrt{\left\{a^2x^2-\left(a^2+b^2\right)\right\}}}{a^2+b^2}$$

Putting this value in (3) and simplifying we get the required locus as

$$a^2x^2 - b^2y^2 = (a^2 + b^2)^2.$$ **Proved.**

Example 17:

Prove that the locus of the pole with respect to the hyperbola $\frac{x^2}{a^2}-\frac{y^2}{b^2}=1$ *of any tangent to the circle whose diameter is the line joining the foci, is the ellipse* $\frac{x^2}{a^4}+\frac{y^2}{b^4}=\frac{1}{a^2+b^2}$.

Solution:

Let the equation to the hyperbola, be

$$\frac{x^2}{a^2}-\frac{y^2}{b^2}=1 \qquad ...(1)$$

If is foci be S and S', their coordinates are respectively (ae, 0) and (–ae, 0)

Equation to the circles drawn on SS' as diameter will be

$$(x - ae)(x + ae) + (y - 0)(y - 0) = 0$$

$$\Rightarrow \quad x^2 + y^2 = a^2e^2 \qquad ...(2)$$

Its centre is (0, 0) and radius is ae.

Now let (h, k) be the pole of any tangent to (2) with respect to (1), the polar of (h, k) with respect to (1) is,

$$\frac{xh}{a^2}-\frac{yk}{b^2}=1 \quad \text{or} \quad b^2\,x\,h - a^2\,k\,y = a^2b^2 \qquad ...(3)$$

If (3) is the tangent to (2) the distance of the centre of the circle (2) from the line (3) must be equal to the radius of the circle, hence,

$$\frac{a^2\,b^2}{\sqrt{a^4\,h^2+a^4\,k^2}}=a\,e$$

$$\Rightarrow \quad a^2e^2\,(b^4h^2 + a^4k^2) = a^2b^2$$

$$\Rightarrow \quad (a^2 + b^2)(b^4h^2 + a^4k^2) = a^2b^4 \qquad (\because a^2c^2 = a^2 + b^2)$$

Generalising and simplifying, the required locus comes to be

$$\frac{x^2}{a^4}+\frac{y^2}{b^4}=\frac{1}{a^2+b^2}.$$ **Proved.**

Example 18:

Prove that the locus of the poles of normal chords with respect to the hyperbola $\frac{x^2}{a^2}-\frac{y^2}{b^2}=1$ *is the curve*

$$y^2a^6 - x^2b^6 = (a^2 + b^2)^2\,x^2y^2$$

Solution:

Let $P \equiv (a \sec \phi, b \tan \phi)$ be any point on the given hyperbola $\frac{x^2}{a^2}-\frac{y^2}{b^2}=1$ then the equation to the normal at P with respect to the hyperbola will be

$$ax \sin \phi + by = (a^2 + b^2) \tan \phi \quad ...(1)$$

Let (h, k) be the pole of the normal given by (1); then the polar of (h, k) with respect to the given hyperbola will be,

$$\frac{xh}{a^2} - \frac{yk}{b^2} = 1$$

$$\Rightarrow \quad b^2xh - a^2yk = a^2b^2 \quad ...(2)$$

As (1) and (2) represent the same straight lines, hence, comparing the coefficients, we get,

$$\frac{a \sin \phi}{b^2 h} = \frac{b}{-a^2k} = \frac{(a^2 + b^2) \tan \phi}{a^2 b^2}$$

hence $\quad \operatorname{cosec} \phi = -\dfrac{a^3k}{b^3h} \quad ...(3)$

and $\quad \cot \phi = -\dfrac{(a^2 + b^2) k}{b^3} \quad ...(4)$

To get the locus of (h, k), we have to eliminate f from (3) and (4).

By plane trigonometry, we have

$$\operatorname{cosec}^2 \phi = 1 + \cot^2 \phi$$

Putting the values from (3) and (4), we get

$$\frac{a^6 k^2}{b^6 h^2} = 1 + \frac{(a^2 + b^2)^2 k^2}{b^6}$$

$$\Rightarrow \quad a^6k^2 = b^6h^2 + (a^2 + b^2) h^2k^2$$

Generalising for (h, k), we get the required locus as,

$$a^6y^2 - b^6x^2 = (a^2 + b^2)^2 x^2y^2.$$ **Proved.**

Example 19:

Prove that the locus of the intersection of tangents to a hyperbola, which meet at a constant angle β, is the curve,

$$(x^2 + y^2 + b^2 - a^2) = 4 \cot^2 \beta \, (a^2y^2 - b^2x^2 + a^2b^2).$$

Solution:

Let the equation to the hyperbola be,

$$\frac{x^2}{a^2} - \frac{y^2}{b^2} = 1 \quad ...(1)$$

Equation to any tangent to (1) is

$$y = mx + \sqrt{(a^2m^2 - b^2)}$$

If this tangent passes through a point (h, k), then we must have,

$$k = mh + \sqrt{(a^2m^2 - b^2)}$$

$$\Rightarrow \quad m^2 (h^2 - a^2) - 2mhk + (k^2 + b^2) = 0 \qquad ...(2)$$

Equation (2) is the quadratic in m, let m_1 and m_2 be the two roots of this equation then if $m_1 = \tan\theta_1$ and $m_2 = \tan\theta_2$, we have

$$\tan\theta_1 + \tan\theta_2 = \frac{2hk}{h^2 - a^2} \text{ and } \tan\theta_1 \tan\theta_2 = \frac{k^2 + b^2}{h^2 - a^2}$$

and as $(\tan\theta_1 - \tan\theta_2)^2 = (\tan\theta_1 + \tan\theta_2)^2 - 4\tan\theta_1 \tan\theta_2$

hence

$$(\tan\theta_1 - \tan\theta_2)^2 = \frac{4h^2k^2 - 4(k^2 + b^2)(h^2 - a^2)}{(h^2 - a^2)^2}$$

$$= \frac{4(a^2k^2 - b^2h^2 + a^2b^2)}{(h^2 - a^2)^2}$$

If the two tangents meet at angle of β, clearly

$$\beta = (\theta_1 - \theta_2)$$

Hence $\cot\beta = \cot(\theta_1 - \theta_1)$

$$= \frac{1}{\tan(\theta_1 - \theta_2)} = \frac{1 + \tan\theta_1 \tan\theta_2}{\tan\theta_1 - \tan\theta_2}$$

$$\Rightarrow \cot^2\beta = \frac{(1 + \tan\theta_1 \tan\theta_2)^2}{(\tan\theta_1 - \tan\theta_2)} = \frac{(h^2 + k^2 + b^2 - a^2)^2}{4(a^2k^2 - b^2h^2 + a^2b^2)^2}$$

Generalising and simplifying, the required locus is,

$$(x^2 + y^2 + b^2 - a^2)^2 = 4\cot^2 b\,(a^2y^2 - b^2x^2 + a^2b^2)^2.$$

Example 20:

Find the locus of the pole of a chord of the hyperbola which subtends a right angle at (1) the centre (2) the vertex and (3) the focus of the curve.

Solution:

(1) Let the equation of the hyperbola be

$$\frac{x^2}{a^2}-\frac{y^2}{b^2}=1 \qquad ...(1)$$

Let P be any point (h, k). The equation of the polar at P with respect to the hyperbola (1) is give by

$$b^2hx - a^2ky = a^2b^2. \qquad ...(2)$$

The equations to the lines joining the point of intersection of (1) and (2) to the centre (0, 0); i.e. the origin will be obtained by making (1) homogeneous with the help of (2). Hence we get

$$\left(b^2 x^2 - a^2 y^2\right) = a^2 b^2 \left\{\frac{a^2 hx - a^2 ky}{a^2 b^2}\right\}^2$$

$$\Rightarrow \quad (b^2xh - a^2yk)^2 = a^2b^2 (b^2x^2 - a^2y^2)^2$$

If these are at right angles, then the sum of the coefficients of x^2 and y^2 must be equal to zero, hence we get,

$$b^4h^2 + a^4k^2 = a^2b^2 (b^2 - a^2)$$

Generalising the locus of the point (h, k) is

$$b^4x^2 + a^4y^2 = a^2b^2 (b^2 - a^2)$$

(2) The coordinates of the vertex are (a, 0). Transferring the origin to this point, the equation of the hyperbola becomes.

$$\frac{(x+a)^2}{a^2}-\frac{y^2}{b^2}=1$$

$$\Rightarrow \quad \frac{x^2}{a^2}-\frac{y^2}{b^2}=\frac{2x}{a}$$

$$\Rightarrow \quad b^2x^2 - a^2y^2 = - 2ab^2x \qquad ...(3)$$

The equation to the polar of (h, k) i.e. equation no (2) of the previous case becomes,

$$b^2 (x + a) h - a^2yk = a^2b^2$$

$$\Rightarrow \quad b^2hx + ab^2h - a^2yk = a^2b^2$$

$$\Rightarrow \quad b^2hx - a^2yk = a^2b^2 - ab^2h \qquad ...(4)$$

The equation of the lines joining the points of intersection of the hyperbola and the chord to the origin is obtained by making (3) homogeneous with the help of (4). Hence, on simplification, this equation becomes

$$(a^2b^2 - ab^2h) (b^2x^2 - a^2y^2) = - 2ab^2x (b^2xh - a^2yk)$$

$$\Rightarrow \quad (a - h) (b^2x^2) - a^2y^2 = -2x (b^2xh - a^2yk)$$

If they are at right angles, the sum of the coefficients of x^2 and y^2 must be zero, hence,

$$(a - h)(b^2 - a^2) + 2b^2h = 0$$

$$\Rightarrow \quad h(a^2 + b^2) = a(a^2 - b^2)$$

Generalising for (h, k), we get the required locus as

$$x = a\left(\frac{a^2 - b^2}{a^2 + b^2}\right)$$

(3) The coefficients of the focus are (ae, 0). Transferring the origin to this point, the equation to the hyperbola becomes,

$$\frac{(x + ae)^2}{a^2} - \frac{y^2}{b^2} = 1$$

$$\Rightarrow \quad b^2(x^2 + a^2e^2 + 2aex) - a^2y^2 = a^2b^2$$

$$\Rightarrow \quad b^2x^2 - a^2y^2 + 2aeb^2x + a^2b^2(e^2 - 1) = 0$$

$$\Rightarrow \quad b^2x^2 - a^2y^2 + 2aeb^2x + b^4 = 0 \quad ...(5)$$

Equation to the polar of (h, k) i.e. the equation (2) of the first case becomes

$$b^2h(x + ae) - a^2k(y + 0) = a^2b^2$$

$$\Rightarrow \quad b^2hx - a^2ky = b^2a^2 - b^2aeh \quad ...(6)$$

Equations to the lines which join the origin to the points of intersection of (5) and (6) will be obtained as in the other cases by making (5) homogeneous with the help of (6), hence we get on simplification

$$a^2(b^2x^2 - a^2y^2)(a - eh)^2 + 2aex(a - eh)(b^2xh - a^2yk) + (b^2xh - a^2yk)^2 = 0$$

If these are at right angles, then the sum of the coefficients of x^2 and y^2 must be zero, hence

$$a^2(b^2 - a^2)(a - eh)^2 + 2a^2b^2eh(a - eh) + b^4h^2 + a^4k^2 = 0$$

$$\Rightarrow \quad h^2(a^2 + 2b^2) - a^2k^2 - 2a^3eh + a^2(a^2 - b^2) = 0$$

Generalising, we get the required locus as

$$x^2(a^2 + 2b^2) - a^2y^2 - 2a^3ex + a^2(a^2 - b^2) = 0. \quad \textbf{Ans.}$$

Example 21:

Chords of a hyperbola are drawn, all passing through the fixed point (h, k), prove that the locus of their middle points is a hyperbola whose centre is the point (h/2, k/2), and which is similar to either the hyperbola or its conjugate.

Solution:

Let the equation to the hyperbola be

$$\frac{x^2}{a^2}-\frac{y^2}{b^2}=1$$

$$\Rightarrow \quad b^2x^2 - a^2y^2 = a^2b^2 \qquad ...(1)$$

The equation to the chord whose middle point is (x_1, y_1), is

$$xx_1b^2 - a^2yy_1 - a^2b^2 = x_1^2 b^2 - a^2y_1^2 - a^2b^2 \quad (\text{by } T = S_1)$$

$$\Rightarrow \quad xx_1b^2 - a^2yy_1 = x_1^2 b^2 - a^2y_1^2$$

If this chord passes through a fixed point (h, k), we must have

$$hx_1b^2 - a^2ky_1 = x_1^2b^2 - a^2y_1^2$$

$$\Rightarrow \quad \frac{1}{a^2}\left(x_1^2 - x_1 h\right) - \frac{1}{b^2}\left(y_1^2 - ky_1\right) = 0$$

which can be written as

$$\frac{\left(x_1-\frac{h}{2}\right)^2}{a^2}-\frac{\left(y_1-\frac{k}{2}\right)^2}{b^2}=\frac{1}{4}\left(\frac{h^2}{a^2}-\frac{k^2}{b^2}\right)$$

Generalising for (x_1, y_1), the required locus is

$$\frac{\left(x-\frac{h}{2}\right)^2}{a^2}-\frac{\left(y-\frac{k}{2}\right)^2}{b^2}=\frac{1}{4}\left(\frac{h^2}{a^2}-\frac{k^2}{b^2}\right)$$

Which is clearly a hyperbola whose centre is (h/2, k/2) and which is similar to the given hyperbola or its conjugate according as the right hand side $\left(\frac{h^2}{a^2}-\frac{k^2}{b^2}\right)$ is positive or negative.

Notes:

(1) If the equation of the hyperbola be $\frac{x^2}{a^2}-\frac{y^2}{b^2}=1$, its asymptotes may be given by $y=\pm\frac{b}{a}x$, hence $\frac{x^2}{a^2}-\frac{y^2}{b^2}=0$ represents both asymptotes.

(2) If the equation to the hyperbola be $\frac{x^2}{a^2}-\frac{y^2}{b^2}=1$, then the conjugate hyperbola is $\frac{y^2}{b^2}-\frac{x^2}{a^2}=1$.

Example 22:

From points on the circle $x^2 + y^2 = a^2$, tangents are drawn to the hyperbola $x^2 - y^2 = a^2$, prove that the locus of the middle points of the chords of contact is the curve,

$$(x^2 - y^2)^2 = a^2 (x^2 + y^2).$$

Solution:

The equation to the hyperbola is

$$x^2 - y^2 = a^2 \qquad ...(1)$$

The equation to the chord having any point (h, k) as its middle point is given by

$$T = S_1$$

$$xh - yk - a^2 = h^2 - k^2 - a^2$$

$$\Rightarrow \quad xh - yk = h^2 - k^2 \qquad ...(2)$$

The equation to the circle is given as $x^2 + y^2 = a^2$, hence any point on it may be taken as $(a \cos \theta, a \sin \theta)$.

The equation to the polar of this point $(a \cos \theta, a \sin \theta)$ with respect to the hyperbola is

$$ax \cos \theta - ay \sin \theta = a^2$$

$$\Rightarrow \quad x \cos \theta - y \sin \theta = a \qquad ...(3)$$

According to the given conditions (2) and (3) represent the same line, hence they are identical. So comparing the co-efficients

$$\frac{\cos \theta}{h}=\frac{\sin \theta}{k}=\frac{a}{h^2-k^2}$$

Hence $$\cos \theta = \frac{ah}{h^2-k^2} \qquad ...(4)$$

and $$\sin \theta = \frac{ka}{h^2-k^2} \qquad ...(5)$$

The required locus will be obtained by eliminating q from (4) and (5). As we have $\sin^2 \theta + \cos^2 \theta = 1$ substituting the values and simplifying, we get

$$(h^2 - k^2)^2 = a^2 (h^2 + k^2)$$

Generalising, the required locus is

$$(x^2 + y^2)^2 = a^2 (x^2 + y^2).$$ **Proved.**

Example 23:

Tangents are drawn to a hyperbola from any point on one of the branches of the conjugate hyperbola. How that their chord of contact will touch the other branch of the conjugate hyperbola.

Solution:

Let the equation to the hyperbola be

$$\frac{x^2}{a^2} - \frac{y^2}{b^2} = 1 \qquad ...(1)$$

Hence the conjugate is

$$\frac{y^2}{b^2} - \frac{x^2}{a^2} = 1 \qquad ...(2)$$

Any point on the conjugate hyperbola i.e. (2) may be taken as (a an f, b sec f). Then the equation to the chord of contact of this point with respect to the hyperbola (1) is

$$\frac{x}{a}\tan\phi - \frac{y}{b}\sec\phi = 1 \qquad ...(3)$$

The equation to the tangent at the point (– a tan f, – b sec f) (which lies on the other branch of the conjugate hyperbola) with respect to the conjugate hyperbola i.e. (2) is

$$\frac{y}{b^2}(-b\sec\phi) - \frac{x}{a^2}(-a\tan\phi) = 1$$

$$\Rightarrow \quad \frac{x}{a}\tan\phi - \frac{y}{b}\sec\phi = 1 \qquad ...(4)$$

As (3) and (4) are same. **Hence proved.**

Example 24:

Prove that the chords of a hyperbola, which touch the conjugate hyperbola are bisected at the point of contact.

Solution:

Let the equation to the hyperbola be

$$\frac{x^2}{a^2} - \frac{y^2}{b^2} = 1 \quad ...(1)$$

Its conjugate will be

$$\frac{y^2}{b^2} - \frac{x^2}{a^2} = 1 \quad ...(2)$$

Any point on (2) is (a tan f, be sec f). The equation to the tangent at this point to the curve (2) will be

$$\frac{y}{b} \sec \phi - \frac{x}{a} \tan \phi = 1 \quad ...(3)$$

Equation to the chord of (1) having the point (a tan ϕ, b sec ϕ) as mid point will be

$$\frac{x.a \tan \phi}{a^2} - \frac{y.b \sec \phi}{b^2} = \frac{a^2 \tan^2 \phi}{a^2} - \frac{b^2 \sec^2 \phi}{b^2}$$

$$\Rightarrow \quad \frac{x \tan \phi}{a} - \frac{y \sec \phi}{b} = \tan^2 \phi - \sec^2 \phi = -1$$

$$\Rightarrow \quad \frac{y}{b} \sec \phi - \frac{x}{a} \tan \phi = 1 \quad ...(4)$$

Example 25(a):

Through the positive vertex of the hyperbola a tangent is drawn, where does it meet the conjugate hyperbola ?

Solution:

If the hyperbola be $\frac{x^2}{a^2} - \frac{y^2}{b^2} = 1$, its positive vertex is (a, 0) and the tangent at (a, 0) is

$$x = a \quad ...(1)$$

The equation of the conjugate hyperbola will be

$$\frac{y^2}{b^2} - \frac{x^2}{a^2} = 1 \quad ...(2)$$

Solving (1) and (2) we get $y = \pm b\sqrt{2}$, hence the required points are $(a, \pm b\sqrt{2})$.

Example 25(b):

Find the equation to the hyperbola, whose asymptotes are the straight lines x + 2y + 3 = 0 and 3x + 4y + 5 = 0 and which passes through the point (1, –1).

Write down also the equation of the conjugate hyperbola.

Solution:

Equations to the asymptotes are given as

$$x + 2y + 3 = 0 \qquad ...(1)$$

and $$3x + 4y + 5 = 0 \qquad ...(2)$$

(1) and (2) may be given by

$$(x + 2y + 3)(3x + 4y + 5) = 0 \qquad ...(3)$$

As the equation to the hyperbola will differ from (3) only by a constant, it may be given by

$$(x + 2y + 3)(3x + 4y + y = \lambda \qquad ...(4)$$

(where λ is a constant)

As (1, –1) lies on the hyperbola will be

$$(x + 2y + 3)(3x + 4y + 5) = 8$$

$$\Rightarrow \quad 3x^2 + 10xy + 8y^2 + 14x + 22y + 7 = 0$$

The equation to the conjugate hyperbola may be obtained by replacing λ by $-\lambda$.

Therefore, the equation of the conjugate hyperbola will be

$$(x + 2y + 3)(3x + 4y + 5) = -8$$

$$3x^2 + 10xy + 8y^2 + 14x + 22y + 23 = 0.$$ **Ans.**

Example 26:

In a rectangular hyperbola, prove that CP and CD are equal and are inclined to the axis at angles which are complementary.

Solution:

For any hyperbola $\frac{x^2}{a^2} - \frac{y^2}{b^2} = 1$, if CP and CD be any two conjugate diameters meeting the hyperbola and its conjugate at P and D respectively, then by, we have

$$CP^2 = a^2 \sec^2 \phi + b^2 \tan^2 \phi$$

and $$CD^2 = a^2 \tan^2 \phi + b^2 \sec^2 \phi$$

As for the rectangular hyperbola a = b, hence,

$$CP^2 = a^2 (\sec^2 \phi + \tan^2 \phi) \text{ and } CD^2 = a^2 (\sec^2 \phi + \tan^2 \phi)$$

Hence $CP^2 = CD^2$ or $CP = CD$

Again equations to CP and CD are respectively

$$y = x.\frac{b}{a}\sin\phi \quad \text{and} \quad y = x\frac{b}{a\sin\phi}$$

For a rectangular hyperbola as a = b, the equation to CP and CD become

$$y = x\sin\phi \text{ and } y = \frac{x}{\sin\phi} \text{ respectively}$$

so slope of CP is sin f and that of CD is 1/sin ϕ. Hence the inclination of CP and CD to the transverse axis are complementary.

Example 27:

Find the asymptotes of the curve $2x^2 + 5xy + 2y^2 + 4x + 5y = 0$ and find the general equation of all hyperbolas having the same asymptotes.

Solution:

The hyperbola is given as

$$2x^2 + 5xy + 2y^2 + 4x + 5y = 0 \qquad \text{...(1)}$$

As the equation to the asymptote will differ from the equation of the hyperbola only in constant term, so it may be given by

$$2x^2 + 5xy + 2y^2 + 4x + 5y + \lambda = 0 \qquad \text{...(2)}$$

If (2) represents two straight lines, we must have

$$abc + 2fg - af^2 - bg^2 - ch^2 = 0$$

$$\Rightarrow \quad 2.2.\lambda + 2.\frac{5}{2}.\frac{4}{2}.\frac{5}{2} - 2.\frac{25}{4} - 2.4 - \lambda\frac{25}{4} = 0$$

$$\Rightarrow \quad 16l + 100 - 50 - 32 - 25\lambda = 0$$

$$\Rightarrow \quad 9\lambda = 18 \qquad \therefore \lambda = 2$$

Putting in (2) the equation of asymptotes will be

$$2x^2 + 5xy + 2y^2 + 4x + 5y + 2 = 0$$

$$\Rightarrow \quad (2x + y + 2)(x + 2y + 1) = 0 \qquad \text{...(3)}$$

As the equations of all hyperbolas will differ from (3) only by a constant, hence it may be given by

$$(2x + y + 2)(x + 2y + 1) = \text{constant.}$$

Example 28:

A series of hyperbolas is drawn having a common transverse axis of length 2a, prove that the locus of a point P on each hyperbola, such that *its distance from the transverse axis is equal to its distance from on asymptote, is the curve $(x^2 - y^2)^2 = 4x^2 (x^2 - a^2)$.*

Solution:

Let the equation to the hyperbola be

$$\frac{x^2}{a^2} - \frac{y^2}{b^2} = 1 \text{ or } b2x2 - a^2y^2 = a^2b^2 \quad ...(1)$$

(Hence b is the variable)

Equation to the asymptote to (1) is

$$\frac{x}{a} = \frac{y}{b}$$

or $\quad b\,x - a\,y = 0 \quad ...(2)$

Let P be any point (h, k) on the curve

$$b^2h^2 - a^2k^2 = a^2b^2 \quad ...(3)$$

Distance of (h, k) from the asymptote (2) is

$$\frac{b\,h - a\,k}{\sqrt{a^2 + b^2}} = k$$

(by hypothesis as the perpendicular distance of (h, k) from are transverse axis i.e. x-axis is k)

or $\quad (bh - ak)^2 = k^2 (a^2 + b^2)$

Hence $\quad b = \dfrac{2a\,h\,k}{h^2 - k^2} \quad ...(4)$

The required locus will be obtained by eliminating ϕ from (3) and (4).

Putting the value of b in (3); we get,

$$\frac{4a^2 h^2 k^2}{(h^2 - k^2)^2} h^2 - a^2 k^2 = a^2 \frac{4a^2 h^2 k^2}{(h^2 - k^2)^2}$$

$$\Rightarrow \quad \frac{4a^2 h^2 k^2}{(h^2 - k^2)^2} \left[h^2 - a^2\right] = a^2 \frac{4a^2 h^2 k^2}{(h^2 - k^2)^2}$$

$$\Rightarrow \quad 4h^2 (h^2 - a^2) = (h^2 - k^2)^2$$

Generalising for (h, k), we get the required locus as

$$4x^2 (x^2 - a^2) = (x^2 - y^2)^2$$

Proved.

Example 29:

C is the centre of the hyperbola $\frac{x^2}{a^2}-\frac{y^2}{b^2}=1$ *and the tangent at any point P meet the asymptotes in the points Q and R. Prove that the equation to the locus of the centre of the circle circumscribing the triangle CQR is* $4(a^2x^2 - b^2y^2) = (a^2 + b^2)^2$.

Solution:

The equation to the hyperbola is given as

$$\frac{x^2}{a^2}-\frac{y^2}{b^2}=1 \quad ...(1)$$

Let P be any point on it as (a sec ϕ, b tan ϕ), then the equation of tangent at P is

$$\frac{x}{a}-\frac{y}{b}\sin\phi=\cos\phi \quad ...(2)$$

The equations to the asymptotes to (1) are,

$$\frac{x}{a}=\frac{y}{b} \quad ...(3)$$

and $$\frac{x}{a}=-\frac{y}{b} \quad ...(4)$$

Solving (2) and (3), we get the coordinates of Q, as

$$\left(\frac{a\cos\phi}{1-\sin\phi},\frac{b\cos\phi}{1-\sin\phi}\right)$$

Solving (2) and (4), we get the coordinates of R, as

$$\left(\frac{a\cos\phi}{1+\sin\phi},\frac{-b\cos\phi}{1-\sin\phi}\right)$$

Let O be the centre of the circle passing through C, Q and R having its coordinates as (h, k). Then clearly OC = OQ

$$\Rightarrow \quad 2(ah+bk)=(a^2+b^2)\frac{\cos\phi}{1-\sin\phi} \quad ...(5)$$

Similarly OC = OR, hence

$$h^2+k^2=\left(h-\frac{a\cos\phi}{1+\sin\phi}\right)^2+\left(k+\frac{b\cos\phi}{1+\sin\phi}\right)^2$$

which on simplification as in last case, gives

$$2\ (a\ h - b\ k) = \left(a^2 + b^2\right) \frac{\cos\phi}{1+\sin\ \phi} \qquad ...(6)$$

To get the locus of the point 'O', we have to eliminate ϕ from (5) and (6), so multiplying the two, we get,

$$4\left(a^2h^2 - b^2k^2\right) = \left(a^2 + b^2\right) \frac{\cos^2\phi}{1-\sin^2\phi} = \left(a^2 + b^2\right)^2$$

Generalising for (h, k) we get the required locus as

$$4\ (a^2x^2 - b^2y^2) = (a^2 + b^2)^2.$$ **Proved.**

Example 30(a):

A straight line always passes through a fixed point, prove that the locus of the middle point of the portion of it, which is intercepted between two given straight lines, is a hyperbola whose asymptotes are parallel to the given lines.

Solution:

Take the two lines as the axis of x and axis of y. Let $\frac{x}{h}+\frac{y}{k}=1$ be the equation of the straight line and the fixed point be (a, b).

$$\frac{a}{h}+\frac{b}{k}=1 \qquad ...(1)$$

Let (x, y) be the coordinates of the mid point of the portion of the st. line intercepted between the axes, then

$$x = \frac{h+0}{2}$$

and $$y = \frac{k+0}{2}$$

or $h = 2x$ and $k = 2y$

Putting these values of h and k in (1), we get the required locus as

$$\frac{a}{2x}+\frac{b}{2y}=1$$

$$ax + by = 2xy$$

or $$2xy - ay - bx = 0$$

which is clearly a hyperbola whose asymptotes are given by

$2xy = 0$ i.e. $x = 0$ and $y = 0$. **Proved.**

Example 30(b):

The coordinates of a point are a tan $(\theta + \alpha)$ and b tan $(\theta + \beta)$ where θ is variable, prove that the locus of the point is a hyperbola.

Solution:

We are given that

$$x = a \tan(\theta + \alpha)$$

and $$y = b \tan(\theta + \beta)$$

$$\Rightarrow \quad \tan^{-1}\frac{x}{a} = \theta + \alpha \quad \text{...(1)}$$

and $$\tan^{-1}\frac{y}{b} = \theta + \beta \quad \text{...(2)}$$

To get the required locus, we have to eliminate θ from (1) and (2).

Hence subtracting (2) from (1), we get

$$\tan^{-1}\left(\frac{x}{a}\right) - \tan^{-1}\left(\frac{y}{b}\right) = \alpha - \beta$$

$$\Rightarrow \quad \tan^{-1}\left(\frac{\frac{x}{a} - \frac{y}{b}}{1 + \frac{xy}{ab}}\right) = \alpha - \beta$$

$$\Rightarrow \quad \frac{\frac{x}{a} - \frac{y}{b}}{1 + \frac{xy}{ab}} = \tan(\alpha - \beta)$$

Simplifying, we get the required locus as

$$xy + ab = (bx - ay)\cot(\alpha - \beta)$$

which is a hyperbola. **Proved.**

Example 31:

Show that two concentric hyperbolas, whose axes meet at angle of 45°, cut orthogonally.

Solution:

Let the equation to the rectangular hyperbola be given as

$$x^2 - y^2 = a^2 \quad \text{...(1)}$$

As the asymptotes of this are the axes of the other and vice-versa, hence the equation of the other hyperbola may be written as

$$xy = c^2 \qquad ...(2)$$

Let (1) and (2) meet at some point whose coordinates are (a sec α, a tan α)

Then the tangent at the point (a sec α, a tan α) to (1) is

$$x - y \sin \alpha = a \cos \alpha \qquad ...(3)$$

and the tangent at the point (a sec α, a tan α) to (2) is

$$y + x \sin \alpha = \frac{2c^2}{a} \cos \alpha \qquad ...(4)$$

Clearly the slopes of the tangents given by (4) and (3) are respectively – sin α and 1/sin α, so their product is

$$-\sin \alpha . \frac{1}{\sin \alpha} = -1$$

Hence the tangents are at right angle.

Example 32:

If the ordinate NP at any point P of an ellipse be produced to Q, so that NQ is equal to the subtangent at P, prove that the locus of Q is hyperbola.

Solution:

Let 'P' be any point (a cos ϕ, b sin ϕ) on the ellipse

$$\frac{x^2}{a^2} + \frac{y^2}{b^2} = 1 \qquad ...(1)$$

Equation to tangent at P with respect to (1) is

$$\frac{x}{a} \cos \phi + \frac{y}{b} \sin \phi = 1 \qquad ...(2)$$

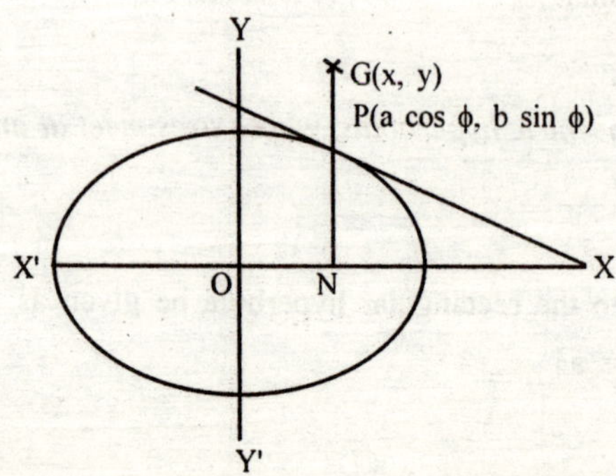

If this tangent meets the axis of x at T, then the abscissa of T will be obtained by putting y = 0 in (2); hence for T.

$$x = a \sec \phi$$

Therefore, the subtangent

$$NT = OT - ON = (a \sec \phi - a \cos \phi)$$

Let PN be the ordinate of P and NP be produced to Q such that NQ = subtangent NT = (a sec ϕ – a cos ϕ).

If the coordinates of Q be (x, y), then clearly

$$x = ON = a \cos \phi \qquad ...(3)$$

$$y = NT = (a \sec \phi - a \cos \phi) \qquad ...(4)$$

The locus of P will be obtained by eliminating f from (3) and (4).

By (3) we get cos ϕ = x/a, now putting this value in (4) have

$$y = a\ (a/x) - a,\ (x/a)$$

or $\quad x^2 + xy = a^2$

which is the required locus. **Proved.**

Example 33:

Prove that the foci of the hyperbola $xy = \dfrac{a^2 + b^2}{4}$ *are given by* $x = y = \pm \dfrac{a^2 + b^2}{2a}$.

Solution:

Let KK' and LL' be the asymptotes which are taken as axes and S be the focus having the coordinates (x, y) with respect to these axes. Draw SH parallel to LL' meeting KK' in H. If 2α be the angle between asymptotes clearly CS will bisect it and hence

$$\angle LCS = \angle SCH = \angle CSH = \alpha$$

Draw HN perpendicular to CS. Then

$$CN = NS = \frac{1}{2} CS$$

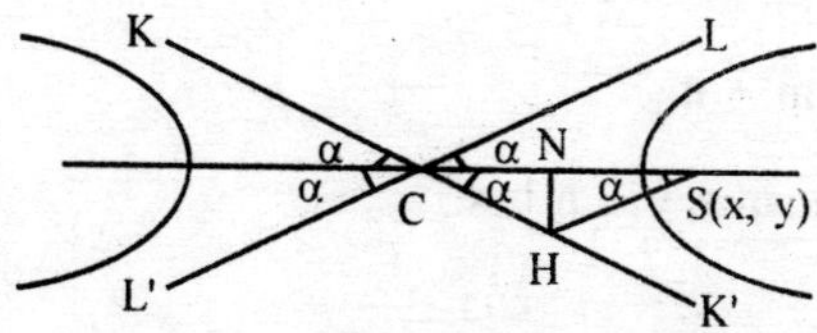

Hence $C\,H \cos\alpha = H\,S \cos\alpha = \frac{1}{2}\,C\,S$

As $\quad CH = x$ and $SH = y$

and $CS = \sqrt{(a^2+b^2)}$, we have

$$x \cos a = y \sin a = \frac{1}{2}\sqrt{a^2+b^2}$$

$$\Rightarrow \qquad x = y = \frac{\sqrt{a^2+b^2}}{2\cos\alpha}$$

also $\quad \tan\alpha = \frac{b}{a} \qquad \therefore \cos\alpha = \frac{\pm a}{\sqrt{a^2+b^2}}$

Hence $\quad x = y = \pm\frac{a^2+b^2}{2a}$. **Proved.**

Example 34:

A variable line has its ends on two lines given in position and passes through a given point; prove that the locus of a point which divides it in any given ratio is a hyperbola.

Solution:

Let the two given lines be the axes and $\frac{x}{h}+\frac{y}{k}=1$ be the equation of the variable line. Since it passes through the fixed point say (a, b) hence

$$\frac{a}{h}+\frac{b}{k}=1 \qquad ...(1)$$

Let the coordinates of the point dividing the line between the axis in the ratio m: n be (x, y), then

$$x=\frac{m}{m+n}\,n \qquad ...(2)$$

and

$$y=\frac{n}{m+n}\,k \qquad ...(3)$$

Eliminating h and k from the relations (1), (2) and (3), we have

$$\frac{ma}{x}+\frac{nb}{y} = m+n$$

or $\quad (m + n)\,x\,y = m\,a\,y + n\,b\,x$

which is a hyperbola. **Proved.**

Example 35:

A series of circles touch a given straight line at a given point. Prove that the locus of the pole of a given straight line with regard to these circles is a hyperbola whose asymptotes are respectively a parallel to the first given straight line and a perpendicular to the second.

Solution:

Let the line which is touched by the circles be taken as y-axis and the point of contact as the origin. Hence the equation to all the circles touching y-axis at origin may be given by

$$x^2 + y^2 = 2\lambda x \qquad ...(1)$$

where λ is the variable.

Again suppose the given straight line is

$$ax + by = c \qquad ...(2)$$

If (h, k) be the pole of (2) w.r.t. (1), then the polar of (h, k) w.r.t. (1) will be

$$xh + yk = \lambda (x + h)$$

$$\Rightarrow \quad x (h - \lambda) + yk = \lambda h \qquad ...(3)$$

As (2) and (3) must be identical, so comparing the coefficients, we get,

$$\frac{h-\lambda}{a} = \frac{k}{b} = \frac{\lambda h}{c} \qquad ...(4)$$

so $$\frac{k}{b} = \frac{\lambda h}{c} \text{ or } \lambda = \frac{ck}{bh} \qquad ...(5)$$

Again by (4), we have

$$\frac{h-\lambda}{a} = \frac{k}{b}$$

Putting the value of λ, from (5), we get

$$\frac{h - \frac{ck}{bh}}{a} = \frac{k}{b} \text{ or } b\,h^2 - c\,k = a\,k\,h$$

Generalising for (h, k) we get the required locus as

$$bx^2 - a\,xy = cy$$

This is a hyperbola, having asymptotes parallel to $x = 0$ and $bx - ay = 0$ i.e. one asymptote parallel to axis of y and other perpendicular to $ax + by = c$. **Hence proved.**

Example 36(a):

From a point P, perpendicular PM and PN are drawn to two straight lines OM and ON. If the area OMPN, be constant, prove that the locus of P is a hyperbola.

Solution:

Let the bisectors of the angle NOM be taken as axes, so the equations of OM and ON are respectively

$$x \cos \alpha + y \sin \alpha = 0$$

and $$x \cos \alpha - y \sin \alpha = 0$$

Take any point P as (h, k), then

PM = perpendicular from P on OM

$$= \frac{h \cos \alpha + k \sin \alpha}{\sqrt{\cos^2 \alpha + \sin^2 \alpha}} = h \cos \alpha + k \sin \alpha \quad ...(1)$$

and similarly PN = perpendicular from P on ON

$$= h \cos \alpha - k \sin \alpha \quad ...(2)$$

$$(h - x) \sin \alpha + (y - k) \cos \alpha = 0$$

$$\Rightarrow \quad y \cos \alpha - x \sin \alpha + h \sin \alpha - k \cos \alpha = 0 \quad ...(3)$$

Similarly the equation to PN will be

$$(h - x) \sin \alpha + (y - k) \cos \alpha = 0$$

$$\Rightarrow \quad -x \sin \alpha - y \cos \alpha + h \sin \alpha + k \cos \alpha = 0 \quad ...(4)$$

Now OM = perpendicular distance of (3) from (0, 0)

$$= \frac{0 - 0 + h \sin \alpha - k \cos \alpha}{\sqrt{\cos^2 \alpha + \sin^2 \alpha}} = h \sin \alpha - k \cos \alpha$$

similarly ON = perpendicular distance of (0, 0) from (4)

$$= h \sin \alpha + k \cos \alpha$$

Now the area of the quad. OMPN = DOMP + DONP

$$= \frac{1}{2} OM \times PM \times \frac{1}{2} ON \times PN$$

$$= \frac{1}{2} [h \sin \alpha - k \cos \alpha] [h \cos \alpha + k \sin \alpha]$$

$$+ \frac{1}{2} [h \sin \alpha + k \cos \alpha] [h \cos \alpha - k \sin \alpha]$$

$= (h^2 - k^2) \sin\alpha \,.\, \cos\alpha = \text{constant} = S \text{ (say)}$

$$\Rightarrow \quad h^2 - k^2 = \frac{S}{\sin\alpha\cos\alpha} = a^2 \quad \text{(say)}$$

which is again constant.

Therefore, the locus of the point (h, k) will be

$x^2 - y^2 = a^2$ which is clearly a hyperbola. **Hence proved.**

Example 36(b):

If a right angled triangle is inscribed in a rectangular hyperbola, prove that the tangent at the right angle is perpendicular upon the hypotenuse.

Solution:

Let the equation to the rectangular hyperbola by $xy = c^2$ and take the coordinates of the vertices of the triangle ABC to be $(ct_1, c/t_1)$, $(ct_2, c/t_2)$ and $(ct_3, c/t_3)$ respectively, $\angle C$ being a right angle.

The gradients of the sides BC and AC are respectively $\frac{-1}{t_1 t_3}$ and $\frac{-1}{t_2 t_3}$.

As the angle ACB is right angle, we must have

$$\frac{-1}{t_1 t_3} \times \frac{-1}{t_2 t_3} = -1$$

$$\Rightarrow \quad t_1 t_2 t_3^2 + 1 = 0 \qquad \text{...(1)}$$

Again the equation of the tangent 't_3' of the given hyperbola will be

$$\frac{x}{t_3} + y\, t_3 = 2c$$

Hence m of the tangent is $-1/t_3^2$

Again m of the line joining A and B will be $-1/t_1 t_2$.

If these lines are perpendicular to each other, we must have

$$\frac{-1}{t_3^2} \times \frac{-1}{t_2 t_1} = -1$$

$$\Rightarrow \quad t_1 t_2 t_3^2 + 1 = 0$$

which is true by relation (1). **Hence proved.**

Example 37:

Prove that any chord of a rectangular hyperbola subtends angles which are equal or supplementary (1) at the ends of a perpendicular chord, and (2) at the ends of any diameter.

Solution:

Take the equation of the rectangular hyperbola as $xy = c^2$ and suppose P, Q, R and S to be the extremities of the chords, having the coordinates as $(ct_1, c/t_1)$, $(ct_2, c/t_2)$, $(ct_3, c/t_3)$ and $(ct_4, c/t_4)$ respectively.

As the chords are given to be perpendicular, we have

$$\frac{-1}{t_1 t_2} \times \frac{-1}{t_3 t_4} = -1 \text{ or } t_1\, t_2\, t_3\, t_4 = -1 \quad ...(1)$$

and the m's of PQ and are $-1/t_1t_2$ and $-1/t_3t_4$ respectively.

$$\text{Hence} \quad \tan PQS = \frac{\frac{-1}{t_1 t_2} + \frac{1}{t_2 t_4}}{1 + \frac{1}{t_1 t_2^2 t_4^2}} = \frac{t_2(t_1 - t_4)}{1 + t_1 t_2^2 t_4} \quad ...(2)$$

By (1), we have $t_1t_2t_4 = -1/t_3$

Putting in (2), we get

$$\tan PQS = \frac{t_2 t_1 - t_2 t_4}{1 - \frac{t_2}{t_3}} = \frac{t_2 t_3 (t_1 - t_4)}{(t_3 - t_2)} \quad ...(3)$$

$$\text{Similarly} \tan PRS = \pm \left\{ \frac{t_2 t_3 (t_1 - t_4)}{(t_3 - t_2)} \right\}.$$

Hence ĐPQS and ĐPRS are either equal or supplementary. Proved (2) Take t_1 and t_2 to the extremities of the chord PQ and let t and – t be the extremities of the diameter RR',

$$\tan PRQ = \frac{t(t_2 \sim t_1)}{1 + t_1 t_2 t^2} \text{ and } \tan PR'Q = \frac{t(t_2 \sim t_1)}{1 + t_1 t_2 t^2}$$

Therefore, the angles PRQ and PR'Q are either equal or supplementary.

Example 38:

Find the coordinates of the points of contact of common tangents to the two hyperbolas

$$x^2 - y^2 = 3a^2 \text{ and } xy = 2a^2.$$

Solution:

The two curves are given as

$$x^2 - y^2 = 3a^2$$

and $$xy = 2a^2 \quad ...(2)$$

Suppose a common tangent touches (1) and (2) at A and B respectively. The coordinates of A may be taken as ($\sqrt{3}$ a sec α, $\sqrt{3}$ a tan α), as in lies on (1).

The equation of tangent at this point w.r.t. (1) will be

$$\frac{x}{\sqrt{3a^2}} - \frac{y}{\sqrt{3a^2}} \sin\alpha = \cos\alpha$$

$$\Rightarrow \quad x - y\sin\alpha = \sqrt{(3)}\, ay\cos\alpha - 2a^2 = 0 \quad ...(4)$$

If the two values of this equation coincide, its discriminant must be zero, hence we must have

$$3a^2\cos^2\alpha - 4(-2a^2)\sin\alpha = 0$$

$$\Rightarrow \quad 3\cos^2 a + 8\sin\alpha = 0$$

$$\Rightarrow \quad 3(1 - \sin^2 a) + 8\sin\alpha = 0$$

$$\Rightarrow \quad 3\sin^2\alpha - 8\sin\alpha - 3 = 0$$

$$\Rightarrow \quad (3\sin\alpha + 1)(\sin\alpha - 3) = 0$$

Hence $$\sin\alpha = -1/3 \quad ...(5)$$

(as sin α cannot be equal to 3)

Hence $$\cos\alpha = \pm\frac{2\sqrt{2}}{3} \quad ...(6)$$

Putting these values in (4), we get

$$y^2\left(-\frac{1}{2}\right) \pm \sqrt{3}\, a\, y\frac{2\sqrt{2}}{3} - 2a^2 = 0$$

or $$y^2 \pm 2\sqrt{6}\, ay + 6a^2 = 0$$

whence $y = \pm\sqrt{6}$.

Putting the values of y, sin a and cos α in (3) and simplifying, we get

$$x = \pm\frac{1}{2}\sqrt{6}a$$

Hence the coordinates of the point of contact of the common tangent with (2) are $(\pm 1/3)\ \sqrt{6a}\ +\ \sqrt{6a}\ )$.

Again putting the values of sec α and tan α with the help of relations (5) and (6) for the coordinates of the point A, we find the coordinates as

$$\left(\pm \frac{3\sqrt{6}}{4}a, \pm \frac{\sqrt{6}}{4}a\right)$$

which is the point of contact of the common tangent with (1). **Ans.**

Example 39:

Show that the of any tangent to the rectangular hyperbola $xy = c^2$ with respect to the circle $x^2 + y^2 = a^2$ lies on a concentric and similarly placed rectangular hyperbola.

Solution:

Equation to the hyperbola is given as

$$xy = c^2 \qquad \text{...(1)}$$

and that of the circle is

$$x^2 + y^2 = a^2 \qquad \text{...(2)}$$

Take (h, k) to be pole w.r.t. (2) of any line which is a tangent to (1).

The polar of (h, k) w.r.t. (2) will be

$$hx + ky = a^2 \qquad \text{...(3)}$$

Solving (1) and (3), we get the equation

$$\frac{x}{k}\left(c^2 - hx\right) = c^2$$

$$\Rightarrow \qquad hx^2 - a^2x + kc^2 = 0 \qquad \text{...(4)}$$

If (3) touches (1), then this equation i.e. (4) must be a perfect square i.e. discriminant must be zero, hence

$$hx^2 - a^2x + kc^2 = 0 \qquad \text{...(4)}$$

$$a^4 - 4khc^2 = 0$$

or $$hk = \frac{a^4}{4c^2}$$

Generalising for (h, k), we get $xy = a^4/4c^2$ which is a rectangular hyperbola similar to the given one having the same centre. **Proved.**

Example 40:

If P_1, P_2 and P_3 be three points on the rectangular hyperbola $xy = c^2$, whose abscissae are x_1, x_2 and x_3. Prove that the area of the triangle $P_1 P_2 P_3$ is

$$\frac{c^2}{2}.\frac{(x_2-x_3)(x_3-x_1)(x_1-x_2)}{x_1\ x_2\ x_3}$$

and that the tangents at these points form a triangle whose area is

$$2c^2.\frac{(x_2-x_3)(x_3-x_1)(x_1-x_2)}{(x_2+x_3)(x_3+x_1)(x_1+x_2)}$$

Solution:

Let the equation of the hyperbola by $xy = c^2$.

If P_1 be a point on the hyperbola having its abscissa as x_1, the corresponding value of the ordinate will be given by

$$y_1 = \frac{c^2}{x_1}$$

Hence the coordinates of the point P, will be $(x_1, c^2/x_1)$ similarly the coordinates of P_2 and P_3 will be $(x_2, c^2/x_2)$ and $(x_3, c^2/x_3)$ and hence the area of the $\Delta\ P_1P_2P_3$ will be

$$= \frac{1}{2}\left[\frac{c^2}{x_1}(x_2-x_3)+\frac{c^2}{x^2}(x_3-x_1)+\frac{c^2}{x_3}(x_1-x_2)\right]$$

$$\therefore \Delta P_1P_2P_3 = \frac{c^2}{2}\left[\frac{x_2\,x_3\,(x_2-x_3)+x_1\,x_3\,(x_3-x_1)+x_1\,x_2\,(x_2-x_1)}{x_1\,x_2\,x_3}\right]$$

Simplifying and factorizing the numerator by the method of cyclic order, we get the required area as

$$\frac{c^2}{2}\left[\frac{(x_2-x_3)(x_3-x_1)(x_2-x_3)}{x_1\ x_2\ x_3}\right]$$

Again the equation to the tangent at point P i.e. $(x_1, c^2/x_1)$ w.r.t. to the hyperbola will be

$$\frac{x}{x_1}+\frac{yx_1}{c^2}=2$$

or $\qquad c^2x + yx_3^2\ yx_1^2 = 2c^2x_1 \qquad$ ...(1)

Similarly, the equation to the tangents at

$P_2 \equiv (x_2, c^2/x_2^2)$

and $P_3 \equiv (x_3, c^2/x_3^2)$ will be

$$c^2x + yx_1^2 = 2c^2x_2 \qquad ...(2)$$

$$c^2x + yx_3^2 = 2c^2x_3 \qquad ...(3)$$

Let (1) and (2) meet in A. To get the coordinates of A, we have to solve these two.

Hence subtracting (2) from (1), we have

$$y\left(x_1^2 - x_2^2\right) = 2c^2 (x_1 - x_2)$$

Hence y $[2c^2/(x_1 + x_2)]$. Putting this value in (1) we get,

$$x = \frac{1}{c^2}\left[x^2 x_1 - \frac{2c^2 x_1^2}{x_1 + x_2}\right] = \frac{2x_1 x_2}{x_1 + x_2}$$

Hence, the coordinates of A are $\left(\frac{2x_1 x_2}{x_1 + x_2}, \frac{2c^2}{x_1 + x_2}\right)$

Similarly, if B and C be points of intersection of (2) and (3) and (1) and (2) their coordinates will be respectively.

$$\left(\frac{2x_2 x_3}{x_2 + x_3}, \frac{2c^2}{x_2 + x_3}\right) \text{ and } \left(\frac{2x_1 x_3}{x_1 + x_3}, \frac{2c^2}{x_1 + x_3}\right)$$

Hence the area of triangle ABC is given by

$$\Delta ABC = \frac{1}{2}\left[\frac{2x_1x_2}{x_1 + x_2}\left(\frac{2c^2}{x_2 + x_3} - \frac{2c^2}{x_1 + x_3}\right) + \frac{2x_2 x_3}{x_2 + x_3}\left(\frac{2c^2}{x_1 + x_3} - \frac{2c^2}{x_1 + x_2}\right)\right.$$
$$\left. + \frac{2x_1 x_3}{x_1 + x_3}\left(\frac{2c^2}{x_1 + x_2} - \frac{2c^2}{x_1 + x_3}\right)\right]$$

$$= \frac{1}{2}2c^2\left[\frac{2x_1 x_2}{x_1 + x_2}\left\{\frac{x_1 + x_3 - (x_2 + x_3)}{(x_2 + x_3)(x_1 + x_3)}\right\} + \frac{2x_2 x_3}{x_2 + x_3}\left\{\frac{x_1 + x_2 - (x_1 + x_3)}{(x_1 + x_3)(x_1 + x_2)}\right\}\right.$$
$$\left. + \frac{2x_1 x_3}{x_1 + x_3}\left\{\frac{x_2 + x_3 - (x_1 + x_2)}{(x_2 + x_3)(x_1 + x_2)}\right\}\right]$$

$$= \frac{2c^2\left[x_1 x_2 (x_1 - x_2) + x_2 x_3 (x_2 - x_3) + x_1 x_3 (x_3 - x_1)\right]}{(x_1 + x_2)(x_2 + x_3)(x_3 + x_1)}$$

Simplifying and factorizing the numerator we get the area of the triangle ABC

$$= 2c^2 \frac{(x_1 - x_2)(x_2 - x_3)(x_3 - x_1)}{(x_1 + x_2)(x_2 + x_3)(x_3 + x_1)} \quad \text{(Omitting the – ive sign).}$$

Example 41:

In a rectangular hyperbola, show that the angle between a chord PQ and tangent at P is equal to the angle which PQ subtends at the other end of the diameter through P.

Solution:

Suppose the equation of the rectangular hyperbola be $xy = c^2$ and take the coordinates of P and Q as $(ct_1, c/t_1)$ and $(ct_2, c/t_2)$ respectively.

The slope of the line PQ is clearly $-1/t_1 t_2$.

Let the angle between the line and the tangent be θ, then

$$\tan\theta = \frac{\frac{1}{t_2^1} - \frac{1}{t_1 t_2}}{1 + \frac{1}{t_1^2 t_2}} = \frac{t_1(t_2 - t_1)}{1 + t_1^2 t_2}$$

Again if the angle which PQ subtends at the other end of the diameter which will be $-t_1$ i.e. $(-ct_1, -c/t_1)$ be ϕ, then

similarly $$\tan\phi = \frac{t_1(t_2 - t_1)}{1 + t_1^2 t_2}$$

Hence the angles θ and ϕ are equal. **Proved.**

Example 42(a):

Show that the normal to the rectangular hyperbola $xy = c^2$ at the point 't' meets the curve again at a point 't' such that

$$t^3 t' = -1.$$

Solution:

Let the equation of the hyperbola be $xy = c^2$ and the coordinates of any point P on it be $(ct, c/t)$ then the equation to the normal at P with respect to the hyperbola is

$$y - x\,t^2 = \frac{c}{t}(1 - t^4)$$

$$\Rightarrow \quad yt - xt^2 - c + ct^4 = 0 \qquad \text{...(1)}$$

If it passes through any point say Q having its coordinates as "t'" i.e. (ct', c/t'), the coordinates must satisfy (1).

hence, we get $\frac{c}{t'}t - c\,t'\,t^3 - c + ct^4 = 0$

$$\Rightarrow \quad t - t'^3t^3 - t' + t't^4 = 0$$

Factorizing, we get,

$$(t't^3 + 1)\,(t - t') = 0$$

Hence $t't^3 + 1 = 0$ or $t't^3 = -1$ as $t \neq t$. **Proved.**

Example 42(b):

In a rectangular hyperbola prove that all straight lines, which subtend a right angle at a point P on the curve, are parallel to the normal at P.

Solution:

Take any point 'P' having the coordinates (ct, c/t) on the rectangular hyperbola, whose equation is $xy = c^2$.

If A and B be any two points on the curve such that AB subtends a right angle at 'P' and the coordinates of A and B be respectively $(ct_1, c/t_1)$ and $(ct_2, c/t_2)$, we have

$$\text{the slope of AB} = \frac{-1}{t_1 t_2} = m_1 \text{ (say)} \qquad \text{...(1)}$$

Similarly, the slope of AP is $-1/tt_1$ and that of BP is $-1/tt_2$.

As AP and BP are perpendicular to each other

$$\frac{-1}{t\,t_1} \times \frac{-1}{t\,t_2} = -1$$

$$\text{or} \quad \frac{1}{t_1 t_2\, t^2} = -1 \qquad \text{...(3)}$$

Now the equation to the normal at P is

$$y - t^2x = \frac{c}{t}\left(1 - t^4\right)$$

Hence the slope of the normal is $t^2 = m^2$ say ...(4)

Comparing (3) and (4) we get $m_1 = m_2$ and hence AB is parallel to the normal at P.

Example 42(c):

Show that the straight lines drawn from a variable point on the curve to any two fixed points on it intercept a constant distance on either asymptote.

Solution:

Let the equation to the curve be given as

$$xy = c^2 \qquad ...(1)$$

and A and B be two fixed points on it as $(ct_1, c/t_1)$ and $(ct_2, c/t_2)$. Take the variable point P as (ct, c/t).

The equation to the lines PA and PB will be respectively

$$x + y\, tt_1 = c\,(t + t_1)$$

and $$x + y\, tt_2 = c\,(t + t_2) \qquad ...(3)$$

If (2) and (3) meet the axis of x at K and L respectively, the abscissae of K and L will be obtained by putting y = 0 in (2) and (3) abscissa of K is $c\,(t + t_1)$ and that of L will be $c\,(t + t_2)$.

Therefore, intercept KL = $c\,(t + t_1) - c\,(t_1 + t_2) = c\,(t_1 - t_2)$

which is constant for all values of t.

Similarly, let M and N be the points where (2) and (3) meet y-axis, their coordinates may be obtained by putting x = 0 in (2) and (3).

So, the coordinates of M is $c\left(\frac{1}{t}+\frac{1}{t_1}\right)$ and that of N is $c\left(\frac{1}{t}+\frac{1}{t_2}\right)$

Hence M N = $c\left(\frac{1}{t}+\frac{1}{t_1}\right) - c\left(\frac{1}{t}+\frac{1}{t_2}\right) = c\left(\frac{1}{t_1}-\frac{1}{t_2}\right)$

which is again constant for all values of t. **Proved.**

Example 43:

The transverse axis of a rectangular hyperbola is 2c and the asymptotes are the axes of coordinates. Show that the equation of the chord which is bisected at the point (2c, 3c) is

$$3x + 2y = 12\,c.$$

Solution:

Under the gives conditions, the equation of the hyperbola may be taken as

$$xy = c^2 \qquad ...(1)$$

The chord having (2c, 3c) as its middle point will be parallel to the polar of (2c, 3c) w.r.t. (1). The equation of this polar will be

$$3cx + 2cy = 2c^2$$

Therefore, the slope of the polar is (–3/2) and passing through (2c, 3c) will be

$$(y - 3c) = - 3/2\ (x - 2c)$$

$$\Rightarrow \quad 3x + 2y = 12\,c.$$

Proved.

Example 44:

Chords of a rectangular hyperbola and at right angles and they subtend a right angle at a fixed point 'O', prove that they intersect at the polar of 'O'.

Solution:

Let the fixed point 'O' be (h, k) and the equation to the rectangular hyperbola be $xy = c^2$.

The equation to the polar of (h, k) w.r.t. (1) is

$$xk + yh = 2c^2 \qquad ...(2)$$

Changing the origin to the point (h, k), the equation of the curve becomes

$$(x + h)\ (y + k) = c^2 \qquad ...(3)$$

and the equation to the polar becomes

$$k\ (k + h) + h\ (y + k) = 2c^2$$

$$\Rightarrow \quad kx + hy = 2\ (c^2 - hk) = 2\lambda \text{ (say)} \qquad ...(4)$$

Again suppose $lx + my = p$ and $ly = q$ be two perpendicular chords of (3). Then the equations of the two pairs of lines joining the origin to the extremities of these chords will be

$$xyp^2 + (kx + hy)\ (lx + my)\ p = \lambda\ (lx + my)^2 \qquad ...(5)$$

and $$xyq^2 + (kx + hy)\ (mx - ly)\ q = \lambda\ (mx - ly)^2 \qquad ...(6)$$

As each pair is at right angles, we must have the sum of the coefficients of x^2 and y^2 equal to zero, hence

$$p\ (lk + mh) = \lambda\ (l^2 + m^2) \text{ from (5)}$$

and $$q\ (mk - lh) = \lambda\ (l^2 + m^2) \text{ from (6)}$$

This is same as the condition that the point of intersection of the chords, i.e.

$$\left(\frac{pl + qm}{l^2 + m^2}, \frac{pm - ql}{l^2 + m^2}\right)$$

should be on the polar given by (4). **Proved.**

Example 45:

Prove that the locus of the poles of all normals chords of the rectangular hyperbola $xy = c^2$ is the curve

$$(x^2 - y^2) + 4c^2 xy = 0$$

Solution:

Any normal to the curve $xy = c^2$ at the point $(ct, c/t)$ is given as

$$xt^2 - ty - ct^4 + c = 0 \qquad ...(1)$$

If its pole is (h, k) then the polar of (h, k) w.r.t. $xy = c^2$ is

$$kx + hy = 2c^2 \qquad ...(2)$$

As (1) and (2) represent the same line, comparing, we get

$$\frac{k}{t^2} = \frac{h}{-t} = \frac{2c^2}{-c + ct^4} \qquad ...(3)$$

From the last two of (3), $h = \dfrac{-2ct}{t^4 - 1}$

From first and last of (3) $k = \dfrac{2ct}{t^4 - 1}$

hence $\qquad \dfrac{k}{h} = -t^2 \qquad ...(4)$

Again $\qquad h^2 = \dfrac{4c^2t^2}{(t^4 - 1)^2} = \dfrac{-4c^2 (k/h)}{(k^2/h^2 - 1)^2} \qquad$ from (4)

$\Rightarrow \qquad (k^2 - h^2) + 4c^2hk = 0$

Generalising, the locus of the pole (h, k) will be

$$(y^2 - x^2)^2 + 4c^2xy = 0.$$ **Proved.**

Example 46:

Show that the equation to the director circle of the conic $xy = c^2$ is

$$x^2 + 2xy \cos \omega + y^2 = 4c^2 \cos \omega.$$

Solution:

The equation to the hyperbola is given as

$$xy = c^2 \qquad ...(1)$$

If 't_1' be any point on it, then the equation of the tangent at 't_1' is

$$t_1^2 y + x = 2ct_1 \qquad ...(2)$$

Similarly, the equation of tangent at any other point 't_2' will be

$$t_2^2 y + x = 2ct_2 \qquad ...(3)$$

Solving (2) and (3), we get

$$x = \frac{2ct_1 t_2}{t_1 + t_2} \text{ and } y = \frac{2c}{t_1 + t_2}$$

Hence, $t_1 t_2 = \dfrac{x}{y}$ ...(4)

and $(t_1 + t_2) = \dfrac{2c}{y}$ ...(5)

Now, if the tangents given by (2) and (3) be perpendicular to one another, and ω be the angle at which the axes are inclined, then we must have,

$$1 + \left(\frac{-1}{t_1^2} - \frac{1}{t_2^2}\right)\cos\omega + \frac{1}{t_1^2}\cdot\frac{1}{t_2^2} = 0$$

$$\Rightarrow \quad 1 - \frac{(t_1^2 + t_2^2)\cos\omega}{t_1^2\, t_2^2} + \frac{1}{t_1^2 t_2^2} = 0$$

$$\Rightarrow \quad t_1^2 t_2^2 - [(t_1 + t_2)^2 - 2t_1 t_2]\cot\omega + 1 = 0 \qquad ...(6)$$

Putting the values of $t_1\, t_2$ and $(t_1 + t_2)$ from (4) and (5), we get the locus of point of intersection of two perpendicular tangents

$$\Rightarrow \quad \frac{x^2}{y^2} - \left(\frac{4c^2}{y^2} - \frac{2x}{y}\right)\cos\omega + 1 = 0$$

$$\Rightarrow \quad x^2 - (4c^2 - 2xy)\cos\omega + y^2 = 0$$

$$\Rightarrow \quad x^2 + y^2 + 2xy\cos\omega = 4c^2\cos\omega.$$ **Proved.**

Example 47(a):

If a hyperbola be rectangular, and its equations be $xy = c^2$ *prove that the locus of the middle points of chords of constant 2d is*

$$(x^2 + y^2)(xy - c^2) = d^2 xy.$$

Solution:

Let (h, k) be the coordinates of the mid point of the given chord of the hyperbola $xy = c^2$ and suppose

$$\frac{x - h}{\cos\theta} = \frac{y - k}{\sin\theta} = r \qquad ...(1)$$

to be the equation of any line through the point (h, k).

Putting the values of x and y from (1) in the equation of the hyperbola, we get the points in which this line cuts the hyperbola $xy = c^2$ are given by

$$(h + r\cos\theta)(k + r\sin\theta) = c^2$$

$$\Rightarrow \quad r^2 \sin\theta\cos\theta + r(k\cos\theta + h\sin\theta) + (hk - c^2) = 0$$

As (h, k) is the middle point of this chords, we have

$$k = \cos\theta + h\sin\theta = 0 \qquad \text{...(3)}$$

$$\Rightarrow \quad \frac{\sin\theta}{k} = -\frac{\cos\theta}{h} = \frac{1}{\sqrt{h^2+k^2}} \qquad \text{...(4)}$$

Putting the value from (3) in (2), we get

$$r^2 + \frac{c^2 - hk}{\cos\theta\sin\theta}$$

$$\Rightarrow \quad r = \sqrt{\frac{c^2 - hk}{\cos\theta\sin\theta}} = \text{half the chord} = d \text{ (by hypothesis)}$$

$$\Rightarrow \quad d^2 = \frac{c^2 - hk}{\cos\theta\sin\theta} = \frac{(c^2 - hk)(h^2 + k^2)}{hk} \quad \text{[by relation (4)]}$$

$$\Rightarrow \quad hkd^2 = (hk - c^2)(h^2 + k^2)$$

Generalising the locus of the middle point (h, k) of the chord is

$$(xy + c^2)(x^2 + y^2) = d^2 xy.$$ **Proved.**

Example 47(b):

A tangent to the parabola $x^2 = 4ay$ meets the hyperbola $xy = k^2$ in two points P and Q. Prove that the middle point of PQ lies on a parabola.

Solution:

The equation of the hyperbola is given as

$$xy = k^2 \qquad \text{...(1)}$$

and the equation of the parabola is

$$y^2 = 2ax \qquad \text{...(2)}$$

The (α, β) the coordinates of the middle point of some chord of (1), then the equation to the chord will be

$$x\alpha + x\beta = 2\alpha\beta \qquad \text{...(3)}$$

If (3) is a tangent to (2), then on solving the two, we must get two coincident values. Hence putting the value of x from (3) in (2), we get

$$y^2 = 4a\left(\frac{2\alpha\beta - y\beta}{\alpha}\right)$$

$$\Rightarrow \quad \alpha y^2 + 4\alpha\beta y - 8a\alpha\beta = 0 \qquad ...(4)$$

(4) has equal roots if its discriminant is zero i.e. when

$$16a^2\beta^2 + 4a\ 8a\ \alpha\beta = 0$$

Generalising for a and b, the required locus of the mid point is

$$2x^2 + ay = 0$$

which is a parabola. **Proved.**

Example 47(c):

Prove that the portions of any line which is intercepted between the asymptotes and the curve are equal.

Solution:

Take the curve to be $xy = c^2$ ...(1)

and let AB be any chord of it having (h, k) as its mid point. Then the equation of polar of (h, k) w.r.t. (1) will be

$$kx + hy = 2c^2$$

Its slope is $-k/h$.

Hence the equation of AB, which is parallel to the polar and passes through (h, k), will be

$$(y - k) = -\frac{k}{h}(x - h)$$

or $\quad hy + kx + 2kh \qquad ...(2)$

The equations to the asymptotes of (1), are

$$x = 0 \qquad ...(3)$$

and $\quad y = 0 \qquad ...(4)$

Let the line AB cut the asymptotes (3) and (4) at P and Q respectively, then the coordinates of 'P' will be (0, 2k) and those of Q will be (2h, 0).

Hence the coordinates of the mid point of PQ will be

$$\left(\frac{2h+0}{2}, \frac{Q+2k}{2}\right) \text{ or } (h, k)$$

which are the same as for the mid point of AB.

Hence, the intercept between the asymptotes and the curve must be equal.

Example 48:

A circle passing through the centre of a rectangular hyperbola, cuts the curve in the points A, B, C and D. Prove that the circumcircle of the triangle formed by the tangents at A, B and C goes through the centre of the hyperbola and has the centre at the point of the hyperbola which is diametrically opposite to D.

Solution:

Refer Art. 334 Ex. 2 As the circle passes through the origin, hence $k = 0$ If $k = 0$, then

$$t_1t_2 + t_2t_3 + t_3t_4 + t_1t_3 + t_1t_4 + t_2t_4 = 0$$

Putting the values from equation (2) of the Ex. 2, we get

$$t_1t_2 + t_2t_3 + t_1t_3 + \frac{1}{t_1\, t_2\, t_3}(t_1 + t_2 + t_3) = 0 \qquad ...(1)$$

Let (x, y) be the coordinates of one of the vertices of the triangle formed by the tangents (say at t_1 and t_2), then

$$x = \frac{2c\, t_1\, t_2}{t_1 + t_2} \text{ and } y = \frac{2c}{t_1 + t_2}$$

Substituting in (1) and simplifying, we get

$$x^2 + y^2 + 2c\, t_4\, x + \frac{2c}{t_4} y = 0.$$

This is the equation of a circle passing through the origin and whose centre is the point $(-ct_4, -c/t_4)$ which is diametrically opposite to D.

Example 49:

If from a fixed point on a rectangular hyperbola, perpendiculars are let fall on any two conjugate diameters, the straight line joining the feet of these perpendiculars has a constant direction.

Solution:

Take the conjugate diamter for the hyperbola

$$x^2 - y^2 = a^2 \text{ to be } y = mx \qquad ...(1)$$

and $$y = \frac{1}{m} x \qquad ...(2)$$

The equation of the line passing through the fixed point (a sec f, a tan f) and perpendicular to the diameter y = mx will be

$$y - a \tan \phi = -\frac{1}{m}(x - a \sec \phi) \qquad ...(3)$$

Solving (1) and (3) the coordinates of the foot of the perpendicular say P, are

$$\left(\frac{a\, m \tan \phi + \sec \phi}{1 + m^2}, \frac{a\, m (\tan \phi + \sec \phi)}{1 + m^2}\right)$$

Again the equation of the line through (a sec ϕ, a tan ϕ) and perpendicular to (2) is

$$y - a \tan \phi = -m(x - a \sec \phi) \qquad ...(4)$$

On solving (2) and (4), the coordinates of the foot of the perpendicular say Q are

$$\left[\frac{a\, m (\tan \phi + m \sec \phi)}{1 + m^2}, \frac{a (\tan \phi + m \sec \phi)}{1 + m^2}\right]$$

The slope of the line joining P and Q is

$$= \frac{a(\tan \phi + m \sec \phi) - am(\tan \phi + \sec \phi)}{am(\tan \phi + m \sec \phi) - a(m \tan \phi + \sec \phi)}$$

$$= -\frac{\tan \phi}{\sec \phi} = -\sin \phi$$

As the point is fixed, ϕ is constant and hence sin f is constant, therefore PQ has a constant slope.

Example 50:

Prove that the locus of the foot of the perpendicular, let fall from the centre upon chord of the rectangular hyperbola $xy = c^2$ *which subtends half a right angle at the origin is the curve*

$$r^4 - 2c^2r^2 \sin 2\phi = c^4$$

Solution:

The equation to the rectangular hyperbola is given as

$$xy = c^2 \qquad ...(1)$$

So the polar coordinates of the foot of the perpendicular from the origin on (2) are (p, α).

Now making (1) homogeneous withe help of (2), we get the equation of the line joining the intersection of (1) and (2) the origin as

$$xy = c^2\left[\frac{x\cos\alpha + y\sin\alpha}{p}\right]^2$$

$$\Rightarrow \qquad p^2xy = c^2 (x \cos \alpha + y \sin \alpha)^2$$

$$\Rightarrow c^2x^2 \cos^2 \alpha + c^2 y^2 \sin^2\alpha + 2\left[c^2 \sin\alpha\cos\alpha - \frac{p^2}{2}\right] = 0 \qquad ...(2)$$

If the angle between the lines represented by (3) is 45°, we must have

$$\tan 45 = \frac{2\sqrt{\left(c^2 \sin\alpha\cos\alpha - \frac{p^2}{2}\right)^2 - c^2\sin^2\alpha\cos^2\alpha}}{c^2\sin^2\alpha + c^2\cos^2\alpha} = 0$$

$$\Rightarrow \qquad = \frac{2\sqrt{\frac{p^2}{4} - p^2c^2\sin\alpha\cos\alpha}}{c^2}$$

Squaring and simplifying, we get,

$$c^4 = p^4 - 4p^2c^2 \sin \alpha \cos \alpha$$

Generalising for (p, a) wc get the required locus as

$$r^4 - 2r^2c^2 \sin 2\theta = c^4$$

Example 51(a):

Prove that the asymptotes to the hyperbola xy = hx + ky are x + k and y = h.

Solution:

The curve is given as

$$xy = hx + ky$$

$$\Rightarrow \qquad (x - h)(y - k) = hk$$

Hence the equations to the asymptotes will be

$$x - h = 0 \quad \text{or} \quad y - k = 0$$

$$x = h \quad \text{or} \quad y = k.$$

Proved.

Example 51(b):

Show that the straight line $y = mx + 2c\sqrt{-m}$ *always, touches the hyperbola* $xy = c^2$ *and that the point of contact is* $(c/\sqrt{-m}, \sqrt{-m})$.

Solution:

The curve is given as $xy = c^2$ ...(1)

and the line is given as $y = mx + 2c\sqrt{-m}$ ...(2)

To solve (1) and (2) we put the value of y from (2) in (1).

Hence, we get $x(mx - 2c\sqrt{(-m)} = c^2$

$\Rightarrow \quad mx^2 - 2xc\sqrt{(-m)} - c^2 = 0$

$\Rightarrow \quad -mx^2 - 2cx\sqrt{(-m)} + c^2 = 0$

$\Rightarrow \quad [\sqrt{(-m)}\, x + c]^2 = 0$ ...(3)

As (3) is a perfect square, it proves that (2) cuts (1) at one point, i.e. touches it. Hence (2) always touches (1).

Again by (3) $x = c/\sqrt{(-m)}$. Hence the abscissa of the point of contact is $c/\sqrt{(-m)}$.

Putting in (1), we get

$$y = \frac{c^2}{c/\sqrt{m}} = c\sqrt{(1-m)}$$

Therefore, the coordinates of the point of contact are

$$\left\{\frac{c}{\sqrt{-m}}, c\sqrt{-m}\right\}.$$

Example 51(c):

A point moves on the given straight line $y = mx$, *prove that the locus of the foot of the perpendicular let fall from the point upon its polar with respect to the ellipse* $\frac{x^2}{a^2}+\frac{y^2}{b^2}=1$ *is a rectangular hyperbola one of whose asymptotes is the diameter of the ellipse which is conjugate to the given straight line.*

Solution:

Any point on the given line y = mx may be taken as (h, mh). Its polar with respect to $\frac{x^2}{a^2}+\frac{y^2}{b^2}=1$ will be

$$\frac{hx}{a^2}+\frac{ymh}{b^2}=1 \qquad ...(1)$$

If a line is perpendicular to it, its equation may be written as

$$\frac{mhx}{b^2}-\frac{mh^2}{a^2}=\lambda$$

Substituting for λ (2), we get the equations of the line as,

$$\frac{xmh}{b^2}-\frac{yh}{a^2}=h^2m\left(\frac{1}{b^2}-\frac{1}{a^2}\right)$$

$$\Rightarrow \quad \frac{a^2xm-b^2y}{a^2b^2}=hm\left(\frac{1}{b^2}-\frac{1}{a^2}\right)$$

$$\Rightarrow \quad \frac{a^2xm-b^2y}{a^2b^2}=hm\left(\frac{a^2-b^2}{a^2b^2}\right)$$

$$\Rightarrow \quad hm(a^2-b^2)=a^2xm-b^2y \qquad ...(2a)$$

Again from (1), we have $b^2xh = a^2ymh = a^2b^2$ ...(3)

On dividing (3) by 2a we get

$$\frac{h(b^2x+a^2ym)}{hm(a^2-b^2)}-\frac{a^2b^2}{(a^2xm-b^2y)}$$

$\Rightarrow(b^2y - a^2mx)(b^2x + a^2my) + ma^2b^2(a^2 - b^2) = 0$

which is the required locus. **Proved.**

Example 52:

Prove that triangle can be inscribed in the hyperbola $xy = c^2$ sides touch the parabola $y^2 = 4ax$.

Solution:

Let the triangle be formed by the three tangents

$$y = m_1x + \frac{a}{m_1},$$

$$y = m_2 x + \frac{a}{m_2}$$

and $$y = m_3 x + \frac{a}{m_3}$$

to the parabola $y^2 = 4ax$. Solving the equation two by two, the vertices of the triangle will be

$$\left[\frac{a}{m_1 m_2}, a\left(\frac{1}{m_1}+\frac{1}{m_2}\right)\right], \left[\frac{a}{m_2 m_3}, a\left(\frac{1}{m_2}+\frac{1}{m_3}\right)\right] \text{ and } \left[\frac{1}{m_1 m_3}, a\left(\frac{1}{m_1}+\frac{1}{m_3}\right)\right]$$

respectively. If first point lies on the hyperbola $xy = c^2$, we must have

$$\frac{a}{m_1 m_2}.a\left(\frac{1}{m_1}+\frac{1}{m_2}\right) = c^2$$

$$\Rightarrow \quad \frac{1}{m_1}+\frac{1}{m_2} = \frac{c^2 m_1 m_2}{a^2} \qquad ...(1)$$

Similarly by the two, we get

$$\frac{1}{m_2}+\frac{1}{m_3} = \frac{c^2 m_2 m_3}{a^2} \qquad ...(2)$$

and $$\frac{1}{m_3}+\frac{1}{m_1} = \frac{c^2 m_3 m_1}{a^2} \qquad ...(3)$$

Now, we have to show that there may be infinite solutions for these three equations i.e. they are not independent. Subtracting (2) from (1), we get

$$\frac{1}{m_1}-\frac{1}{m_3} = \frac{c^2 m_2 (m_1 - m_3)}{a^2}$$

$$\Rightarrow \quad \frac{m_3 - m_1}{m_1 m_2} = \frac{c^2 m_2 (m_1 - m_3)}{a^2}$$

$$m_2 = -\frac{a^2}{c^2 m_1 m_3}$$

or $$\frac{1}{m_2} = -\frac{c^2 m_1 m_3}{a^2}$$

Substituting this value in (1), we get

$$\frac{1}{m_1} - \frac{c^2 m_1 m_3}{a^2} = \frac{c^2 m_1}{a^2}\left(-\frac{a^2}{c^2 m_1 m_2}\right)$$

$$\Rightarrow \qquad \frac{1}{m_1}+\frac{1}{m_3}=\frac{c^2\ m_3\ m_1}{a^2}$$

Which is same as (3) hence (3) can be deduced from (1) and (2). Therefore, three equations are not independent.

Example 53:

If a circle and rectangular hyperbola meet in four points P, Q, R and S, show that the orthocentres of the triangles QRS, RSP, SPQ and PQR also lie on a circle.

Prove also that the tangents to the hyperbola at R and S meet in a point which lies on the diameter of the hyperbola which is at right angles to PQ.

Solution:

(1) Let the equation of the hyperbola be $xy = c^2$ and that of the circle be

$$x^2 + y^2 - 2hx - 2ky = \lambda = 0.$$

If P, Q, R and S be the points of intersection of the two curves having the coordinates given by $(ct_1, c/t_1)$, r being equal to 1, 2, 3 and 4 respectively, then as in the previous question, we have

$$t_1 + t_2 + t_3 + t_4 = \frac{2h}{c} \qquad \text{...(1)}$$

$$\Sigma t_1\, t_2 = \frac{\lambda}{c^2} \qquad \text{...(2)}$$

$$\Sigma\, t_1\, t_2\, t_3 = 2.\frac{k}{c} \qquad \text{...(3)}$$

and $$t_1t_2t_3t_4 = 1 \qquad \text{...(4)}$$

Now the equation to QR will be $x + yt_2t_3 = c\ (t_3 + t_4)$.

Equation to the line perpendicular to QR and passing through P say PN will be

$$y-\frac{c}{t_1} = t_2t_3\ (x - ct_1)$$

$$\Rightarrow \qquad y + ct_1\, t_2\, t_3 = t_2\, t_3 \left(x+\frac{c}{t_1\, t_2\, t_3}\right) \qquad \text{...(5)}$$

Again similarly the equation to the straight line through Q perpendicular to RP, say QM will be

$$y = ct_1\, t_2\, t_3 = t_3\, t_1 \left(x+\frac{c}{t_1\, t_2\, t_3}\right) \qquad \text{...(6)}$$

Solving (5) and (6), the coordinates of the orthocentre of the triangle PQR will be

$$\left(\frac{-c}{t_1 t_2 t_3}, -c t_1 t_2 t_3\right) \text{ i.e. } \left(-ct_4, \frac{-c}{t_4}\right)$$

Similarly orthocentres of the other triangles, are

$$\left(-ct_1, \frac{-c}{t_1}\right), \left(-ct_2, \frac{-c}{t_2}\right) \text{ and } \left(-ct_3, \frac{-c}{t_3}\right)$$

Clearly, these points lie on a circle as the product of their 't'

$$= (-t_1)(-t_2)(-t_3)(-t_4)\, t_1 t_2 t_3 t_4 = 1$$

(2) The equation of the tangents at the points

$$R \equiv \left(c t_3, \frac{c}{t_3}\right) \text{ and } S \equiv \left(c t_4, \frac{c}{t_4}\right) \text{ will be}$$

$$\frac{x}{t_3} + y\, t_3 = 2c \text{ and } \frac{x}{t_4} + y\, t_4 = 2c \text{ respectively.}$$

Solving the two, the coordinates of the point of intersection of these line (say z), will be

$$\left(\frac{2c\, t_3 t_4}{t_3 + t_4}, \frac{2c}{t_3 + t_4}\right)$$

The equation to PQ will be $x + yt_1t_2 = c\,(t_1 + t_2)$, hence the equation of the diamter perpendicular to AB, i.e. the line passing through (0, 0) and perpendicular to AB, will be

$$y - t_1t_2x = 0 \quad \text{or} \quad y - \frac{x}{t_3t_4} = 0 \left[\because \quad t_1 t_2 = \frac{1}{t_3 t_4} \text{ by (4)}\right]$$

Substituting the coordinates of Z in L.H.S. of this equation, we find

$$\frac{2c}{t_3 + t_4} - \frac{1}{t_3\, t_4} \cdot \frac{2c\, t_3\, t_4}{t_3 + t_4} = 0$$

Hence, the coordinates satisfy the equations. **Hence proved.**

Example 54:

A quadrilateral circumscribes a hyperbola, prove that the straight line joining the middle points of its diagonals passing through the centre of the curve.

Solution:

Take the equation of the hyperbola as $xy = c^2$, and 't' i.e. $(ct_1, c/t_1)$ be the coordinates of the points where r = 1, 2, 3, and 4 at which the tangents are drawn to form the quadrilateral.

The equations of the tangents at these points may be given by $x/t_1 + yt_1 = 2c$ whence r = 1, 2, 3 and 4.

Solving these equation the coordinates of the angular points of the quadrilateral ABCD are respectively

$$\left(\frac{2ct_1 t_2}{t_1+t_2}, \frac{2c}{t_1+t_2}\right), \left(\frac{2ct_2 t_3}{t_2+t_3}, \frac{2c}{t_2+t_3}\right)$$

$$\left(\frac{2ct_3 t_4}{t_3+t_4}, \frac{2c}{t_3+t_4}\right) \text{ and } \left(\frac{2ct_1 t_4}{t_1+t_4}, \frac{2c}{t_1+t_4}\right)$$

Let $P \equiv (\alpha, \beta)$ be the mid point of the diagonal AC, then

$$\alpha = c\left(\frac{t_1 t_2}{t_1+t_2} + \frac{t_3 t_4}{t_3+t_4}\right) = c\frac{\Sigma t_1 t_2 t_3}{(t_1+t_2)(t_3+t_4)}$$

and $$\beta = c\left(\frac{1}{t_1+t_2} + \frac{1}{t_3+t_4}\right) = c\frac{\Sigma t_1}{(t_1+t_2)(t_3+t_4)}$$

$$\therefore \quad \frac{\alpha}{\beta} = \frac{\Sigma t_1 t_2 t_3}{\Sigma t_1}$$

Again if (α', β') be the coordinates of the mid point of the diagonal BD, then,

$$\alpha, = c\left(\frac{t_2 t_3}{t_2+t_3} + \frac{t_1 t_4}{t_1+t_4}\right) = c\frac{\Sigma t_1 t_2 t_3}{(t_2+t_3)(t_1+t_4)}$$

and $$\beta, = c\left(\frac{1}{t_2+t_3} + \frac{1}{t_1+t_4}\right) = c\frac{\Sigma t_1}{(t_2+t_3)(t_1+t_4)}$$

$$\therefore \quad \frac{\alpha'}{\beta'} = \frac{\Sigma t_1 t_2 t_3}{\Sigma t_1} = \frac{\alpha}{\beta} \text{ or } \alpha'\beta = \beta'\alpha \qquad ...(1)$$

The equation to the line joining (a, b) and (a', b') will be

$$y - \beta = \frac{\beta' - \beta'}{\alpha' - \alpha}(x - \alpha)$$

If the line passes through (0, 0), we have

$$-\beta = \frac{\beta'-\beta'}{\alpha'-\alpha}(-\alpha)$$

$$\Rightarrow \quad \alpha'\beta - \alpha\beta = \alpha\beta - \alpha\beta$$

$$\Rightarrow \quad \alpha'\beta = \beta'\alpha \qquad \text{...(2)}$$

which is same as (1). **Hence proved.**

Example 55:

Given five points on a circle of radius a, prove that he centres of the conjugate hyperbolas each passing through four of these points, all be on a circle of radius a/2.

Solution:

Take ($a \cos \alpha_n$, $a \sin \alpha_n$) to be the coordinates of the five points where r = 1, 2, 3, 4 and 5.

Again by Ex. 2, ARt. 334, we see that the centre of mean position of any points is the middle point of the line joining the centre (0, 0) to the centre of the corresponding hyperbolas. Hence if (h, k) be the centre of the hyperbola, through the first four points

$$h = \frac{a}{2}\left(\Sigma \cos a_r - \cos a_5\right) \quad (r = 1, 2, 3 \text{ and } 4)$$

and

$$k = \frac{a}{2}\left(\Sigma \sin a_r - \sin a_5\right) (r = 1, 2, 3 \text{ and } 4)$$

$$\Rightarrow \quad \left(h - \frac{a}{2}\Sigma \cos a_r\right)^2 + \left(k - \frac{a}{2}\Sigma \sin a_r\right)^2 = \frac{a^2}{4}$$

Similar in the case for the other four centres.

Hence the required locus is the circle

$$\left(x - \frac{a}{2}\Sigma \cos a_r\right)^2 + \left(y - \frac{a}{2}\Sigma \sin a_r\right)^2 = \frac{a^2}{4}$$

Its radius is clearly a/2.

Example 56:

A, B, C and D are the points of intersection of a circle and a rectangular hyperbola. If AB passes through the centre of the hyperbola, prove that CD passes through the centre of the circle.

Solution:

Let the equation of the hyperbola be $xy = c^2$ and P be any point on it as $(ct, c/t)$.

Let the equation to the circle be

$$x^2 + y^2 - 2hx - 2ky + \lambda = 0 \qquad ...(1)$$

so that its centre Q is the point (h, k).

Now if P lies on (1), we have

$$c^2 t^2 + \frac{c^2}{t^2} - 2h\,c\,t - 2k\frac{c}{t}t + 1 = 0$$

If the roots of this equation be t_1, t_2, t_3 and t_4, then we have

$$t_1 + t_2 + t_3 + t_4 = \frac{2h}{c} \qquad ...(2)$$

$$t_1 t_2 + t_1 t_3 + t_1 t_4 + t_2 t_3 + t_2 t_4 + t_3 t_4 = \frac{\lambda}{c^2} \qquad ...(3)$$

$$t_2 t_3 t_4 + t_3 t_4 t_1 + t_4 t_1 t_2 + t_1 t_2 t_3 + \frac{2k}{c} \qquad ...(4)$$

and $$t_1 t_2 t_3 t_4 = 1 \qquad ...(5)$$

If A, B, C and D are points of intersection of the circle and the hyperbola, there coordinates may be given by

$$\left(c\,t_1, \frac{c}{t_1}\right), \left(c\,t_2, \frac{c}{t_2}\right), \left(c t_3, \frac{c}{t_3}\right) \text{ and } \left(c\,t_4, \frac{c}{t_4}\right)$$

Equation to AB will be respectively

$$x + y\,t_1 t_2 = c\,(t_1 + t_2) \qquad ...(6)$$

Since this passes through the centre of the hyperbola (0, 0) so, we have

$$c\,(t_1 + t_2) = 0 \text{ or } t_1 + t_2 = 0 \qquad ...(7)$$

Similarly equation to CD will be

$$x + y t_3 t_4 = c\,(t_3 + t_4)$$

If this passes through the centre of the circle (h, k) then

$$h + k t_3 t_4 - c\,(t_3 + t_4) = 0 \qquad ...(8)$$

from (2) and (7) $$t_3 + t_4 = \frac{2h}{c} \qquad ...(9)$$

And from (4) $$t_3\,t_4\,(t_1 + t_2) + t_1\,t_2\,(t_4 + t_3) = \frac{2k}{c}$$

Putting the values from (7) and (9)

$$0 + t_1 t_2 \frac{2h}{c} = \frac{2k}{c} \qquad \therefore t_1 t_2 = \frac{k}{h}.$$

Example 57:

If a rectangular hyperbola circumscribes a triangle, show that it meets the circle circumscribing the triangle in a fourth point which is at the other end of the diameter of the hyperbola which passes through the orthocentres of the triangle.

Hence prove that the locus of the centre of the hyperbola which circumscribes a triangle is the nine point circle of the triangle.

Solution:

Let $t_1t_2t_3$ be the angular points of the triangle, then the abscissa of the orthocentre will be $(-c/t_1t_2t_3)$ by Ex. 1 of Art 336 and by Art 334 Ex. 2, $c/t_1t_2t_3$ is the fourth point in which the hyperbola meets the circumcircle. Hence, the centre of the hyperbola is midway between these two points.

Again as the middle point of the straight line joining the orthocentre of a triangle to any point on its circumcircle lies on the nine point circle to the triangle.

Again from (5), we have $t_3 t_4 = \dfrac{1}{t_1 t_2} = \dfrac{h}{k}$ ...(10)

Putting the values from (9) and (10) in (8) we have

$$h + k \cdot \frac{h}{k} - c \cdot \frac{2h}{c} = 0$$

Which is identically true. Hence CD passes through the centre (h, k) of the circle.

Example 58:

A series of hyperbolas is drawn, having for asymptotes the principal axes of an ellipse, show that the common chord of the hyperbolas and the ellipse are all parallel to one of the conjugate diameters of the ellipse.

Solution:

Let the equation of the ellipse be

$$\frac{x^2}{a^2} + \frac{y^2}{b^2} = 1 \qquad ...(1)$$

So that the axes are axis of x and axis of y. The equation of the hyperbola having these axis as asymptotes will be

$$xy = c^2 \qquad ...(2)$$

where c is a variable.

The equation of any curve passing through the points of intersection of (1) and (2) may be given by

$$\left(\frac{x^2}{a^2}+\frac{y^2}{b^2}-1\right)+2\lambda\left(xy-c^2\right)=0 \qquad ...(3)$$

If (3) represents two straight lines, its discriminant must be zero, hence

$$ab = h^2 \qquad [\therefore\ g = f = 0 \text{ from } (2)]$$

i.e. $$\frac{1}{a^2\,b^2}=\lambda^2.$$

or $$\lambda=\frac{1}{ab}$$

Putting in (3), we get the equations of the common chords

$$\Rightarrow \quad \left(\frac{x^2}{a^2}+\frac{y^2}{b^2}-1\right)+\frac{2}{ab}\left(xy-c^2\right)=0$$

$$\Rightarrow \quad \left(\frac{x}{a}+\frac{y}{b}\right)^2=1+\frac{2c^2}{ab}$$

$$\Rightarrow \quad \frac{x}{a}+\frac{y}{b}=\pm\sqrt{\left(1+\frac{2c^2}{ab}\right)} \qquad ...(4)$$

The lines represented by (4) are always parallel to $\frac{x}{a}+\frac{y}{b}=0$ for all values of c.

Example 59:

Show that the chord, which joints the points in which a pair of conjugate diameters meets the hyperbola and its conjugate, is parallel to one asymptote and is bisected by the other.

Solution:

If P and Q be the points where the pair of conjugate diameters meet the hyperbola and the conjugate respectively, then if the coordinates of P be (a sec ϕ, b tan ϕ) they by the same article, the coordinates of D will be (a tan ϕ, b sec ϕ)

$$\text{slope of PC} = \frac{b \sec \phi - b \tan \phi}{a \tan \phi - a \sec \phi} = -\frac{b}{a}$$

which is the same of the asymptotes

$$y = -\frac{b}{a}$$

Hence PD is parallel to this asymptote.

Again the middle point of PD is

$$\frac{1}{2} a (\sec \phi + \tan \phi), \frac{1}{2} b (\sec \phi + \tan \phi)$$

This point satisfies the equation to the other asymptote y + (b/a) x, hence lies on it.

Example 60:

If e and e' be the eccentricities of a hyperbola, and its conjugate, prove that

$$\frac{1}{e^2} + \frac{1}{(e)^2} = 1.$$

Solution:

Let $\frac{x^2}{a^2} - \frac{y^2}{b^2} = 1$ be any hyperbola, then

$$e^2 = \frac{a^2 + b^2}{a^2}$$

The equation to its conjugate will be

$$\frac{y^2}{b^2} - \frac{x^2}{a^2} = 1$$

if e' be its eccentricity then

$$(e)^2 = \frac{a^2 + b^2}{b^2}$$

$$\therefore \quad \frac{1}{e^2} + \frac{1}{(e)^2} = \frac{a^2}{a^2 + b^2} + \frac{b^2}{a^2 + b^2} = \frac{a^2 + b^2}{a^2 + b^2} = 1$$

i.e. $$\frac{1}{e^2} + \frac{1}{(e)^2} = 1.$$ **Proved.**

Example 61:

A straight line is drawn parallel to the conjugate axis of a hyperbola to meet it and the conjugate hyperbola in the points P and Q show that the tangents at P and Q meet on the curve.

$$\frac{y^4}{b^4}\left(\frac{y^2}{b^2}-\frac{x^2}{a^2}\right)=\frac{4x^2}{a^2}$$

and that the normals meet on the axes of x.

Solution:

Let the equation to the hyperbola be

$$\frac{x^2}{a^2}-\frac{y^2}{b^2}=1 \qquad ...(1)$$

and its conjugate hyperbola be

$$\frac{y^2}{b^2}-\frac{x^2}{a^2}=1 \qquad ...(2)$$

Let P be any point (a sec ϕ, b tan ϕ) on (1). The equation to the line parallel to the conjugate axis of (1) i.e. y-axis through P will be

$$x = a \sec \phi \qquad ...(3)$$

The line (3) will cut the conjugate hyperbola (2) at Q when x = a sec f and hence $y = b\sqrt{(1+\sec^2\phi)}$. Therefore, the coordinate of Q will be [a sec ϕ, b $\sqrt{(1+\sec^2\phi)}$].

Now, the equation to the tangent to (1) at P is

$$\frac{x}{a}-\frac{y}{b}\sin\phi=\cos\phi$$

$$\Rightarrow \qquad \frac{x}{a}-\cos\phi=\frac{y}{b}\sin\phi \qquad ...(4)$$

and the equation to the tangent to (2) at Q is

$$\frac{y}{b}\sqrt{1+\sec^2\phi}-\frac{x}{a}\sec\phi=-1$$

$$\Rightarrow \qquad \frac{x}{a}+\cos\phi=\frac{y}{b}\sqrt{1+\cos^2\phi} \qquad ...(5)$$

On squaring and adding (4) and (5), we have

$$2\frac{x^2}{a^2}+2\cos^2\phi=\frac{y^2}{b^2}\left[\left(1+\cos^2\phi\right)+\sin^2\phi\right]=\frac{2y^2}{b^2}$$

$$\Rightarrow \qquad \cos^2\phi=\frac{y^2}{b^2}-\frac{x^2}{a^2}$$

Putting the value of cos ϕ in (5), we get

$$\frac{x}{a}+\sqrt{\frac{y^2}{b^2}-\frac{x^2}{a^2}}=\frac{y}{b}\sqrt{1+\frac{y^2}{b^2}-\frac{x^2}{a^2}}$$

squaring, we get

$$\frac{x^2}{a^2}+\frac{y^2}{b^2}-\frac{x^2}{a^2}+2.\frac{x}{a}\sqrt{\frac{y^2}{b^2}-\frac{x^2}{a^2}}=\frac{y^2}{b^2}\left(1+\frac{y^2}{b^2}-\frac{x^2}{a^2}\right)$$

$$\Rightarrow \quad \frac{y^2}{b^2}\sqrt{\frac{y^2}{b^2}-\frac{x^2}{a^2}}=\frac{2x}{a}$$

$$\Rightarrow \quad \frac{y^2}{b^2}\left(\frac{y^2}{b^2}-\frac{x^2}{a^2}\right)=\frac{4x^2}{a^2} \quad \text{(squaring both sides)}$$

Again, the equations to the normals at point P and Q are

$$ax \sin \phi + by = (a^2 + b^2) \tan \phi \qquad ...(6)$$

$$\text{and } ax \sqrt{(1+\cos^2 \phi)} + by = (a^2 + b^2) \left[\sqrt{(1+\sec^2 \phi)}\right] \qquad ...(7)$$

Solving (6) and (7) we get

$$x=\frac{a^2+b^2}{a} \sec \phi \text{ and } y = 0$$

Hence the two normals meet on the line y = 0 i.e. x-axis.

Example 62:

From a point G on the transverse axis GL, is drawn perpendicular to the asymptote, and GP a normal to the curve at P. Prove that LP is parallel to the conjugate axis.

Solution:

Let the equation to the hyperbola be

$$\frac{x^2}{a^2}-\frac{y^2}{b^2}=1 \qquad ...(1)$$

and 'P' be any point on it as (a sec fϕ b tan ϕ).

Then the equation to the normal at P will be

$$ax \,.\, \sin \phi + by = (a^2 + b^2) \tan \phi \qquad ...(2)$$

Let (2) meet the transverse axis i.e. y = 0 at G, then the coordinates of G are $[(a^2 + b^2/a) \sec \phi, 0]$.

The equation of an asymptote is bx – ay = 0 ...(3)

Hence the equation to any perpendicular to (3) will be

$$ax + by = \lambda \quad ...(4)$$

As (4) passes through G ° $\left(\frac{a^2+b^2}{a}\sec\phi, 0\right)$, we have

$$l = (a^2 + b^2) \sec \phi$$

Therefore, the equation of the perpendicular line i.e. (4) becomes

$$ax + by = (a^2 + b^2) \sec \phi \quad ...(5)$$

Solving (2) and (5) for x, we have, x = a sec ϕ.

Hence the abscissa of L is a sec ϕ and that of P is also a sec f, therefore, clearly PL is given by x = a sec ϕ, which is perpendicular to x-axis i.e. perpendicular to transverse axis.

Example 63:

An ellipse and a hyperbola have the same principal axes. Show that the polar of any point on either curves with respect to the other touches the first curve.

Solution:

Let equation of the hyperbola be

$$\frac{x^2}{a^2} - \frac{y^2}{b^2} = 1$$

and that the ellipse be $(x^2/a^2) + (y^2/b^2) = 1$. Take any point P on the hyperbola as (a sec ϕ, b tan ϕ) and the point Q on the ellipse as

(a cos ϕ, b sin ϕ).

Now, we have to prove that the polar of the point P with respect to ellipse touches the given hyperbola and the polar of Q with respect to the hyperbola touches the given ellipse.

The polar of any point P with respect to the ellipse will be

$$\frac{x}{a^2} a \sec \phi + \frac{y}{b^2} b \tan \phi = 1$$

$$\Rightarrow \quad \frac{x}{a} \sec \phi + \frac{y}{b} b \tan \phi = 1 \quad ...(1)$$

Clearly (1) is the equation to the tangent to the hyperbola at the point (a sec ϕ, 2 b tan ϕ).

Again the polar of the point Q with respect to the hyperbola will be

$$\frac{x}{a^2} a \cos \phi + \frac{y}{b^2} b \sin \phi = 1$$

$$\Rightarrow \quad \frac{x}{a} \cos \phi + \frac{y}{b} \sin \phi = 1 \quad \text{...(2)}$$

Again (2) is clearly the equation to the tangent to the given ellipse at the point (a cos ϕ – h sin ϕ).

Example 64:

Prove that the straight line $\frac{x}{a} - \frac{y}{b} = m$ and $\frac{x}{a} + \frac{y}{b} = \frac{1}{m}$ *always meet on the hyperbola.*

Solution:

The equations are given as

$$\frac{x}{a} - \frac{y}{b} = m \quad \text{...(1)}$$

and

$$\frac{x}{a} + \frac{y}{b} = \frac{1}{m} \quad \text{...(2)}$$

To find out the locus of the point of intersection of (1) and (2), we have to eliminate the variable m from these two. So multiplying (1) and (2), we get,

$$\left(\frac{x}{a} - \frac{y}{b}\right)\left(\frac{x}{a} + \frac{y}{b}\right) = m.\frac{1}{m}$$

or

$$\frac{x^2}{a^2} - \frac{y^2}{b^2} = 1$$

which is clearly a hyperbola.

Example 65:

Find the equation to the chord of the hyperbola

$$25x^2 - 16y^2 = 400$$

which is bisected at the point (5, 3).

Solution:

The hyperbola is given as $25x^2 - 16y^2 = 400$.

The point is given as (5, 3).

The equation of the chord which is bisected at (5, 3) is given b ...(1)

where $T = 25\ 5x - 16\ 3y - 400 = 125x - 48y - 400$

and $S_1 = 25\ (5)^2 - 16\ (3)^2 - 400 = (615 - 144 - 400)$

Putting in (1), we get the required equation as

$$125x - 48y - 400 = 625 - 144 - 400$$

$$\Rightarrow \quad 125x - 48y = 481.$$

Example 66:

Through the positive vertex of the hyperbola a tangent is drawn, where does it meet the conjugate hyperbola ?

Solution:

If the hyperbola be $\frac{x^2}{a^2} - \frac{y^2}{b^2} = 1$, its positive vertex is (a, 0) and the tangent at (a, 0) is

$$x = a \qquad \text{...(1)}$$

The equation of the conjugate hyperbola will be

$$\frac{y^2}{b^2} - \frac{x^2}{a^2} = 1 \qquad \text{...(2)}$$

Solving (1) and (2) we get $y = \pm b\sqrt{2}$, hence the required points are $(a, \pm b\sqrt{2})$.